Computational Mathematics in Engineering

Computational Mathematics in Engineering

S.A. Hovanessian

Lexington Books
D.C. Heath and Company
Lexington, Massachusetts
Toronto

Library of Congress Cataloging in Publication Data
Hovanessian, Shahen A. 1931-
Computational mathematics in engineering.
Includes index.
1. Engineering mathematics. I. Title.
TA330.H68 519.4'02'462 76-14667
ISBN 0-669-00733-1

Published simultaneously in Canada

Printed in the United States of America

International Standard Book Number: 0-669-00733-1

Library of Congress Catalog Card Number: 76-14667

To the memory of my teachers

Louis A. Pipes

Professor of Engineering

and

Ivan S. Sokolnikoff

Professor of Mathematics

University of California, Los Angeles

Contents

List of Figures

List of Tables

Preface

Over the past several years, the computational methods available to engineers have greatly expanded. This, together with the fact that a greater number of engineers use digital computers, has made it desirable to compile the numerical methods applicable to the solution of engineering problems under one cover.

This text is written on a senior or graduate level for students of engineering and practicing engineers. The basics of numerical analysis and computational methods are covered in the first three chapters. These chapters consist of numerical evaluation of matrices and simultaneous equations, calculus of finite differences, and numerical solution of differential equation.

Chapters 4 and 5 cover the basics of probability theory and the least squares methods and estimates. Chapter 4 gives a brief review of probability methods and theorems which are applicable to the digital simulation of probabilistic problems. In addition, this chapter prepares the reader for the discussion of least squares methods and estimates given in the next chapter. Chapter 5 starts with a discussion of classical least squares polynomial fits and proceeds to cover recursive trackers, linear least square methods, and Kalman filtering.

Chapters 6, 7, and 8 discuss various numerical methods of special interest. Chapter 6 gives a discussion of Fourier series and transforms and follows this with computational methods and algorithms. Chapter 7 covers some of the currently used optimization methods, such as linear programming, quadratic programming, and nonlinear programming methods. Chapter 8 gives several numerical methods for the computation of characteristic values and vectors.

In the use of the book as a text, the first three chapters are basic to a knowledge of numerical methods and should be covered. Additional material from Chapters 4 through 8 may be included, depending on the orientation of the class. For example, for students of electrical engineering, the material of Chapters 4, 5, and 6 is desirable, while for students of civil and mechanical engineering, Chapters 6, 7 and 8 may be appropriate.

Each chapter contains about twenty problems, with answers to every problem given at the end of the book. A solution book will also be provided by the author to instructors upon written request.

S.A. Hovanessian

Los Angeles, California
April 1976

Computational Mathematics in Engineering

1 Numerical Evaluation of Matrices and Simultaneous Equations

Matrices form an important branch of mathematics, as they are the logical extension of vectors. Vectors can be considered as one-dimensional matrices while matrices, in their usual representation, are considered as two dimensional. Higher dimension (more than two) matrices are replaced by tensors. Thus, vectors and matrices can be considered one- and two-dimensional tensors, respectively.

In applied mathematics matrices are encountered in the solution of simultaneous equations, numerical solution of boundary value differential equation, numerical solution of partial differential equations, and in a number of problems in electrical circuits and mechanical vibrations. The formulation of many engineering problems for numerical solution results in a set of simultaneous linear equations which can be solved using matrix methods and digital computers.

This chapter discusses some useful properties of matrices and will give several numerical methods for the evaluation of matrices and simultaneous linear equations.

Some Matrix Properties

1. Multiplication of some row i of an nth order matrix by a constant α is equivalent to premultiplication of the matrix by

$$\begin{matrix} \\ \\ \\ \\ i \\ \\ \\ \end{matrix}\begin{pmatrix} & & & i & & & \\ 1 & & & & & & \\ & \ddots & & \vdots & & & \\ & & 1 & & & & \\ \cdot & \cdot & \cdot & \alpha & & & \\ & & & & 1 & & \\ & & & & & \ddots & \\ & & & & & & 1 \end{pmatrix} \qquad (1.1)$$

where the above matrix is an nth order unity matrix with the element of the ith row, ith column replaced by α. As an example, consider the multiplication of the second row of the third-order matrix

$$\begin{pmatrix} a & b & c \\ x & y & z \\ u & v & w \end{pmatrix} \tag{1.2}$$

by α. This can be written as

$$\begin{pmatrix} 1 & 0 & 0 \\ 0 & \alpha & 0 \\ 0 & 0 & 1 \end{pmatrix}\begin{pmatrix} a & b & c \\ x & y & z \\ u & v & w \end{pmatrix} = \begin{pmatrix} a & b & c \\ \alpha x & \alpha y & \alpha z \\ u & v & w \end{pmatrix} \tag{1.3}$$

2. Consider an nth order matrix. Adding to the elements of some row (i) numbers proportional to the elements of some preceding row (j) is equivalent to the premultiplication of the matrix by

$$\begin{array}{cc}
 & \begin{array}{cccccc} & & j & & & \end{array} \\
\begin{array}{c} \\ \\ \\ \\ i \\ \\ \\ \end{array} & \begin{pmatrix} & 1 & \vdots & & & \\ & & \ddots & & & \\ & & 1 & & & \\ & & \vdots & \ddots & & \\ \cdots & & \alpha & \cdots & 1 & \\ & & & & \ddots & \\ & & & & & 1 \end{pmatrix}
\end{array} \tag{1.4}$$

where the above matrix represents an nth order unity matrix with the element of the ith row and jth column replaced by the proportionality constant α. As an example, consider the third-order matrix of equation (1.2). Let it be required to multiply the elements of the second row of this matrix by α and add these to the elements of the third row. This can be accomplished as follows:

$$\begin{pmatrix} 1 & 0 & 0 \\ 0 & 1 & 0 \\ 0 & \alpha & 1 \end{pmatrix} \begin{pmatrix} a & b & c \\ x & y & z \\ u & v & w \end{pmatrix} = \begin{pmatrix} a & b & c \\ x & y & z \\ y+\alpha x & v+\alpha y & w+\alpha z \end{pmatrix} \quad (1.5)$$

3. Consider an nth order matrix. Addition to the elements of some row (i) numbers proportional to the elements of some following row (j) is equivalent to premultiplication of the matrix by

$$\begin{pmatrix} 1 & & & \vdots & & & \\ & \ddots & & & & & \\ & & 1 & \cdots \; \alpha \; \cdots & & \cdots & \\ & & & \ddots \;\; \vdots & & & \\ & & & 1 & & & \\ & & & & \ddots & & \\ & & & & & & 1 \end{pmatrix} \begin{matrix} \\ \\ i \\ \\ \\ \\ \\ \end{matrix} \quad (1.6)$$

with j marking the column of α,

where the above matrix represents an nth order unity matrix with the element in row (i) column (j) replaced by the proportionality constant α. As an example, consider the multiplication of the third row of the matrix of equation (1.2) by α and the subsequent addition of this row to the second row. This can be accomplished by matrix multiplication

$$\begin{pmatrix} 1 & 0 & 0 \\ 0 & 1 & \alpha \\ 0 & 0 & 1 \end{pmatrix} \begin{pmatrix} a & b & c \\ x & y & z \\ u & v & w \end{pmatrix} = \begin{pmatrix} a & b & c \\ x+\alpha u & y+\alpha v & z+\alpha w \\ u & v & w \end{pmatrix} \quad (1.7)$$

Upper and Lower Triangular Matrices

Consider the following third-order triangular matrices

$$\begin{pmatrix} c_{11} & 0 & 0 \\ c_{21} & c_{22} & 0 \\ c_{31} & c_{32} & c_{33} \end{pmatrix} \begin{pmatrix} b_{11} & b_{12} & b_{13} \\ 0 & b_{22} & b_{23} \\ 0 & 0 & b_{33} \end{pmatrix} \tag{1.8}$$

lower triangular matrix   upper triangular matrix

where the lower triangular matrix is defined as a matrix with zero elements above the diagonal and the upper triangular matrix is defined as a matrix with zero elements below the diagonal. With these definitions, we consider the following methods based on the utilization of triangular matrices. These methods, for ease of understanding, are described using third-order matrices. The methods, of course, can be generalized to higher order matrices.

Theorem: On the condition that the leading submatrices of the matrix

$$\mathbf{A} = \begin{pmatrix} a_{11} & a_{12} & a_{13} \\ a_{21} & a_{22} & a_{23} \\ a_{31} & a_{32} & a_{33} \end{pmatrix} \tag{1.9}$$

are nonsingular, i.e.,

$$a_{11} \neq 0, \quad \begin{bmatrix} a_{11} & a_{12} \\ a_{21} & a_{22} \end{bmatrix} \neq 0, \ \ldots, \ [\mathbf{A}] \neq 0 \tag{1.10}$$

matrix **A** may be represented by the product of lower and upper triangular matrices. In equation (1.10), brackets represent determinants of the leading submatrices of (1.9). These submatrices start with the element of row 1 column 1 and proceed to the second- and third-order matrices along the diagonal of the original matrix **A**.

Denoting the lower and upper triangular matrices of **A** by **L** and **U**, respectively, we have

$$\mathbf{A} = \mathbf{LU}$$

$$\mathbf{A} = (a_{ij}) \quad \mathbf{L} = (l_{ij}) \quad \mathbf{U} = (u_{ij}) \tag{1.11}$$

$$a_{ij} = \sum_{k=1}^{\min(i,j)} l_{ik}\, u_{kj}$$

In the above notation a_{ij} represents the element of row i column j of matrix **A**. The lowercase letters are used to denote matrix elements and uppercase letters are used to denote matrices. Note that we will have n^2

equations in $n^2 + n$ unknowns u_{ij} and l_{ij}. Our latitude lies in the specifying of the diagonal coefficients l_{ii} or u_{ii} as seen in the example below.

Example: For a 3 $\times$ 3 matrix we have

$$\begin{pmatrix} a_{11} & a_{12} & a_{13} \\ a_{21} & a_{22} & a_{23} \\ a_{31} & a_{32} & a_{33} \end{pmatrix} = \begin{pmatrix} l_{11} & 0 & 0 \\ l_{21} & l_{22} & 0 \\ l_{31} & l_{32} & l_{33} \end{pmatrix} \begin{pmatrix} u_{11} & u_{12} & u_{13} \\ 0 & u_{22} & u_{23} \\ 0 & 0 & u_{33} \end{pmatrix} \tag{1.12}$$

where the **A** matrix is represented as the product of the lower and the upper triangular matrices **L** and **U**. Performing the matrix product of (1.12) and equating the resulting matrices term by term, we get the set of equations

$$\begin{aligned} a_{11} &= l_{11}\,u_{11} & \qquad a_{21} &= l_{21}\,u_{11} \\ a_{12} &= l_{11}\,u_{12} & \qquad a_{22} &= l_{21}\,u_{12} + l_{22}\,u_{22} \\ a_{13} &= l_{11}\,u_{13} & \qquad a_{23} &= \;\ldots \end{aligned} \tag{1.13}$$

By selecting values of l_{11}, l_{22} and l_{33}, the rest of the values of the triangular matrices can be computed.

Furthermore, the inverse of lower and upper triangular matrices have the same form. For example, the inverse of a lower triangular matrix will also be a lower triangular matrix, as shown below for a third-order matrix

$$\begin{pmatrix} l_{11} & 0 & 0 \\ l_{21} & l_{22} & 0 \\ l_{31} & l_{32} & l_{33} \end{pmatrix} \begin{pmatrix} x_{11} & 0 & 0 \\ x_{21} & x_{22} & 0 \\ x_{31} & x_{32} & x_{33} \end{pmatrix} = \begin{pmatrix} 1 & 0 & 0 \\ 0 & 1 & 0 \\ 0 & 0 & 1 \end{pmatrix} \tag{1.14}$$

Assuming that we already have calculated the elements $l_{11}, l_{21}, \ldots$ of the lower triangular matrix, we can perform the above matrix multiplication and obtain a set of equations for the calculation of the elements of the inverse matrix $x_{11}, x_{21}, x_{22}, \ldots$ This results in

$$\left\{ \begin{aligned} & l_{11}x_{11} = 1 \\ & l_{21}x_{11} + l_{22}x_{21} = 0 \\ & l_{31}x_{11} + l_{32}x_{21} + l_{33}x_{31} = 0 \end{aligned} \right. \tag{1.15}$$

$$\left\{ \begin{aligned} & l_{22}x_{22} = 1 \\ & l_{32}x_{22} + l_{33}x_{32} = 0 \qquad (1.16) \\ & l_{33}x_{33} = 1 \qquad (1.17) \end{aligned} \right.$$

Equations (1.15-17) can be solved for the elements of the inverse matrix. Similarly, for an upper triangular matrix **U** and its inverse **Y** we have

$$\begin{pmatrix} u_{11} & u_{12} & u_{13} \\ 0 & u_{22} & u_{23} \\ 0 & 0 & u_{33} \end{pmatrix} \begin{pmatrix} y_{11} & y_{12} & y_{13} \\ 0 & y_{22} & y_{23} \\ 0 & 0 & y_{33} \end{pmatrix} = \begin{pmatrix} 1 & 0 & 0 \\ 0 & 1 & 0 \\ 0 & 0 & 1 \end{pmatrix} \tag{1.18}$$

Performing the above matrix multiplication will result in the set of simultaneous equations

$$\begin{cases} u_{11}y_{11} = 1 \\ u_{11}y_{12} + u_{12}y_{22} = 0 \\ u_{11}y_{13} + u_{12}y_{23} + u_{13}y_{33} = 0 \end{cases} \tag{1.19}$$

and so on. Assuming we already have the elements u_{11}, u_{12}, u_{12}, . . . we can use equations (1.19) to calculate the values of the inverse matrix y_{11}, y_{12}, . . .

The above methods are used as a means of matrix inversion, as given in the following example.

Example: Consider the matrix

$$\mathbf{A} = \begin{pmatrix} 50 & 107 & 36 \\ 25 & 54 & 20 \\ 31 & 66 & 21 \end{pmatrix} = \mathbf{LU} \tag{1.20}$$

Select diagonal elements of **L** as $l_{11} = 5$, $l_{22} = 1$, and $l_{33} = 0.2$. Using the above relations, we get

$$\mathbf{L} = \begin{pmatrix} 5 & 0 & 0 \\ 2.5 & 1 & 0 \\ 3.1 & -0.68 & 0.2 \end{pmatrix}; \quad \mathbf{U} = \begin{pmatrix} 10 & 21.4 & 7.4 \\ 0 & 0.5 & 2 \\ 0 & 0 & 0.2 \end{pmatrix} \tag{1.21}$$

We also can calculate the elements of the inverse matrices and obtain

$$\mathbf{L}^{-1} = \begin{pmatrix} 0.2 & 0 & 0 \\ -0.5 & 1 & 0 \\ -4.8 & 3.4 & 5 \end{pmatrix} \tag{1.22}$$

and

$$\mathbf{U}^{-1} = \begin{pmatrix} 0.1 & -4.28 & 39.2 \\ 0 & 2 & -20 \\ 0 & 0 & 5 \end{pmatrix} \tag{1.23}$$

Then using the relationship for the inverse of product matrices given below

$$\mathbf{D} = \mathbf{BC};\ \mathbf{D}^{-1} = (\mathbf{BC})^{-1} \tag{1.24}$$

$$(\mathbf{BC})^{-1} = \mathbf{C}^{-1}\mathbf{B}^{-1} \text{ (reverse order)} \tag{1.25}$$

where **B**, **C**, and **D** are square matrices, in equation (1.20) we get

$$\mathbf{A}^{-1} = \mathbf{U}^{-1}\mathbf{L}^{-1} = \begin{pmatrix} -186 & 129 & 196 \\ 95 & -66 & -100 \\ -24 & 17 & 25 \end{pmatrix} \tag{1.26}$$

Choleski's Method

Consider the case where the given matrix **A**, for which an inverse is desired, can be represented as the product of two transposed triangular matrices

$$\mathbf{A} = \mathbf{T}'\mathbf{T} \tag{1.27}$$

where **T** is an upper triangular matrix and **T**′ is its transpose. The transpose of matrix **T** with elements t_{ij} is obtained by transposing matrix elements t_{ij} symmetrically with respect to the diagonal, i.e., elements of transposed matrix **T**′ will be $t'_{ij} = t_{ji}$. Thus, if **T** is an upper triangular matrix, **T**′ will be a lower triangular matrix. The equation for a 2×2 matrix becomes

$$\begin{pmatrix} a_{11} & a_{12} \\ a_{21} & a_{22} \end{pmatrix} = \begin{pmatrix} t_{11} & 0 \\ t_{12} & t_{22} \end{pmatrix} \begin{pmatrix} t_{11} & t_{12} \\ 0 & t_{22} \end{pmatrix} \tag{1.28}$$

Multiplying the matrices of the right side and equating corresponding elements with those of matrix **A**, we get

$$\begin{aligned} a_{11} &= t_{11}^2;\ a_{12} = t_{11}t_{12};\ a_{21} = t_{21}t_{11} \\ a_{22} &= t_{12}^2 + t_{22}^2 \end{aligned} \tag{1.29}$$

Having the values of the elements of matrix **A**, the above equations will give the values of the elements of **T** matrix and subsequently the transposed matrix **T**′ can be defined.

Note that whenever **A** is a symmetrical matrix, i.e., $a_{ij} = a_{ji}$, matrix **A** can be represented as the product of two triangular matrices as given in (1.28). Equations (1.29) can be generalized for an nth order matrix as follows,

$$t_{11} = \sqrt{a_{11}};\ t_{1i} = a_{1i}/t_{11}$$

$$t_{22} = \sqrt{a_{22} - t_{12}^2};\ t_{2i} = (a_{2i} - t_{12}t_{1i})/t_{22}$$

$$\cdots$$

$$t_{nn} = \sqrt{a_{nn} - t_{1n}^2 - \ldots - t_{n-1,n}^2} \tag{1.30}$$

where positive roots are selected after taking the square roots. In the carrying out of the above procedure some of the roots may come out to be imaginary. Although this will make the calculations somewhat cumbersome, it will not affect the values of the final result which are the elements of matrix **T**. After having determined matrix **T**, we can proceed to compute the inverse matrix as shown in the examples below.

Example: Consider the following 2×2 symmetrical matrix

$$\mathbf{A} = \begin{pmatrix} 1 & 1 \\ 1 & 2 \end{pmatrix} \tag{1.31}$$

It is desired to decompose matrix **A** into the product of two triangular symmetric matrices. Matrix **A** can be written as the product of two triangular matrices,

$$\begin{pmatrix} 1 & 1 \\ 1 & 2 \end{pmatrix} = \begin{pmatrix} t_{11} & 0 \\ t_{12} & t_{22} \end{pmatrix} \begin{pmatrix} t_{11} & t_{12} \\ 0 & t_{22} \end{pmatrix} \tag{1.32}$$

Using relations (1.29), we get $t_{11} = 1$, $t_{12} = 1$, $t_{22} = 1$, and (1.32) becomes

$$\begin{pmatrix} 1 & 2 \\ 1 & 2 \end{pmatrix} = \begin{pmatrix} 1 & 0 \\ 1 & 1 \end{pmatrix} \begin{pmatrix} 1 & 1 \\ 0 & 1 \end{pmatrix} \tag{1.33}$$

Inversion Example: Consider the case where a real symmetric matrix is written in the form $\mathbf{A} = \mathbf{LL}'$, where **L** is a lower triangular matrix and prime denotes transpose. It is required to obtain the inverse of matrix **A**. Using the reversal law of the inverse of product matrices, we have

$$\mathbf{A}^{-1} = (\mathbf{LL}')^{-1} = (\mathbf{L}')^{-1}(\mathbf{L})^{-1} \tag{1.34}$$

We also know $(\mathbf{L}')^{-1} = (\mathbf{L}^{-1})'$, resulting in

$$\mathbf{A}^{-1} = (\mathbf{L}^{-1})'(\mathbf{L}^{-1}) \tag{1.35}$$

Given the real valued symmetric matrix **A**,

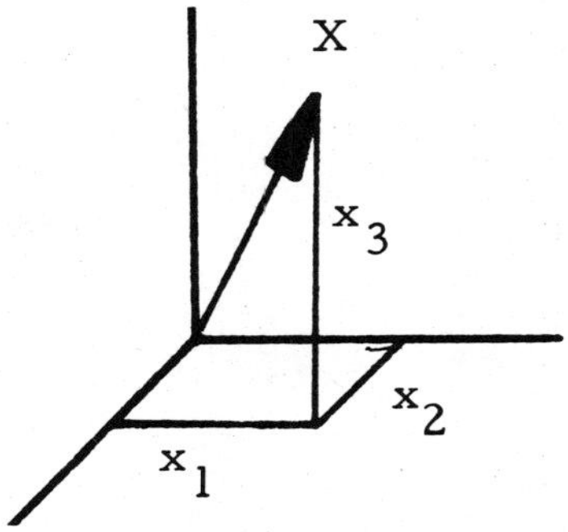

Figure 1-1. Geometric Representation of the Three-Dimensional Vector **X**

$$\mathbf{A} = \begin{pmatrix} 1 & 2 & 6 \\ 2 & 5 & 15 \\ 6 & 15 & 46 \end{pmatrix} \quad \text{we get} \quad \mathbf{L} = \begin{pmatrix} 1 & 0 & 0 \\ 2 & 1 & 0 \\ 6 & 3 & 1 \end{pmatrix} \tag{1.36}$$

and

$$\mathbf{L}^{-1} = \begin{pmatrix} 1 & 0 & 0 \\ -2 & 1 & 0 \\ 0 & -3 & 1 \end{pmatrix} \tag{1.37}$$

resulting in

$$\mathbf{A}^{-1} = (\mathbf{L}^{-1})'(\mathbf{L}^{-1}) = \begin{pmatrix} 5 & -2 & 0 \\ -2 & 10 & -3 \\ 0 & -3 & 1 \end{pmatrix} \tag{1.38}$$

Vectors and Matrices

Consider the three-dimensional vector **X** with its coordinates x_1, x_2, x_3 in a Cartesian coordinate system. The matrix representation and the geometric representation of this vector (Figure 1-1) follow.

$$\mathbf{X} = \begin{pmatrix} x_1 \\ x_2 \\ x_3 \end{pmatrix} \tag{1.39}$$

Certain useful characteristics of vectors dealing with their magnitude, or the magnitude of their components, can be singled out in describing the vector in general terms. Some of these characteristics are categorized and are commonly referred to as *norms* of the vector,

$$
\begin{aligned}
&1.\ \|\mathbf{X}\| = \max_i |x_i| \\
&2.\ \|\mathbf{X}\| = |x_1| + |x_2| + \ldots + |x_n| \qquad (1.40) \\
&3.\ \|\mathbf{X}\| = \sqrt{x_1^2 + x_2^2 + \ldots + x_n^2}
\end{aligned}
$$

The first norm defines the absolute value of the largest component of the n-dimensional vector **X**. The second norm gives the sum of the absolute values of the components of **X**, while the third norm gives the magnitude of the vector. These norms, and/or their combinations, can be extended to matrices. In addition, a number of other norms dealing with characteristic values of matrices are used. Some of these matrix norms will be discussed later in this chapter.

Solution of Simultaneous Equations

Linear simultaneous equations and matrices are directly related, as simultaneous equations can be written in matrix notation. With this commonality most of matrix evaluation methods can also be used for the solution of simultaneous linear equations.

The numerical solutions of simultaneous equations described in the following pages can be divided into two broad categories—iterative solutions and direct solutions. Iterative solutions start with an initial assumed solution and improve this solution on each iteration. The iterations eventually converge to the final solution. In a number of cases, however, depending on the norms of matrices involved, the iterations may not converge to a solution resulting in an unstable condition. The number of iterations involved for convergence to a final solution, in additional to the initial assumed solution, will depend on the relative values of matrix elements involved (norms), as will be discussed later.

Direct methods of solution *do not* involve iterations. These solutions consist of a set of rules which perform arithmetical operations on the matrix elements involved. These operations are directed towards an eventual solution. In direct methods the total number of arithmetical operations (addition, multiplication, and so on) involved for a complete solution can be determined beforehand. This number depends on the order of the matrix or the number of simultaneous equations. In iterative methods, although the number of arithmetical operations per iteration can be calculated, the total number of operations cannot be estimated as the number of iterations is an unknown.

Iterative Solution of Gauss

Assume we have a set of three equations in three unknowns x_1, x_2, x_3 in the form

$$\begin{aligned} x_1 &= a_{11}x_1 + a_{12}x_2 + a_{13}x_3 + f_1 \\ x_2 &= a_{21}x_1 + a_{22}x_2 + a_{23}x_3 + f_2 \\ x_3 &= a_{31}x_1 + a_{32}x_2 + a_{33}x_3 + f_3 \end{aligned} \tag{1.41}$$

The above is not the common form for a set of three equations and three unknowns. Equations (1.41) show the set of equations after solving for, x_1, x_2, and x_3, respectively, in terms of other unknowns and the constants. Note that a_{11}, a_{22}, a_{33} may be zero. In matrix notation, (1.41) becomes

$$\mathbf{X} = \mathbf{AX} + \mathbf{F} \tag{1.42}$$

where **X** is the column vector of the unknowns, **A** is the matrix of the coefficients, and **F** is the column vector of constants. Consider the initial iteration value of $\mathbf{X}^0$ and denote the succeeding iterations by superscripts 1, 2, . . ., and so on. Using this, we may write.

$$\begin{aligned} \mathbf{X}^1 &= \mathbf{AX}^0 + \mathbf{F} \\ \mathbf{X}^2 &= \mathbf{AX}^1 + \mathbf{F} \\ &\vdots \\ \mathbf{X}^k &= \mathbf{A}^k\mathbf{X}^0 + (\mathbf{I} + \mathbf{A} + \mathbf{A}^2 + \ldots \mathbf{A}^{k-1})\mathbf{F} \end{aligned} \tag{1.43}$$

Using these notations, the following convergence theorems apply.

Theorem: Convergence of the iteration process with any arbitrary initial vector $\mathbf{X}^0$ will occur if the characteristic values of matrix **A** are less than unity. This theorem ensures that the series $\mathbf{I} + \mathbf{A} + \mathbf{A}^2 + \ldots$ within the parentheses of (1.43) converges.

The characteristic values of **A** are obtained by setting the determinant $|\mathbf{A} - \lambda\mathbf{I}| = 0$, where λ is the characteristic value and **I** is the unity matrix. Equating this determinant to zero will result in a polynomial of order n in λ where n is the order of matrix **A**. Note that the calculation of characteristic values for large matrices is a formidable task, as will be seen in Chapter 8. In practice, it is sometimes easier to try a solution for convergence on the digital computer than to compute the characteristic values.

Theorem: It is also *sufficient* for the process of iteration to converge under the condition that any norm of matrix **A** be less than unity. Some matrix norms are given at the end of the next section.

Preparation for Iteration Process

Consider the set of equations given in matrix notation $\mathbf{BX} = \mathbf{G}$. For a set of three unknowns x_1, x_x, and x_3, this equation can be written in expanded form,

$$\begin{cases} b_{11}x_1 + b_{12}x_2 + b_{13}x_3 = g_1 \\ b_{21}x_1 + b_{22}x_2 + b_{23}x_3 = g_2 \\ b_{31}x_1 + b_{32}x_2 + b_{33}x_3 = g_3 \end{cases} \tag{1.44}$$

These equations can be written in the form,

$$\begin{cases} x_1 + \dfrac{b_{12}}{b_{11}} x_2 + \dfrac{b_{13}}{b_{11}} x_3 = \dfrac{g_1}{b_{11}} \\[2ex] \dfrac{b_{21}}{b_{22}} x_1 + x_2 + \dfrac{b_{23}}{b_{22}} x_3 = \dfrac{g_2}{b_{22}} \\[2ex] \dfrac{b_{31}}{b_{33}} x_1 + \dfrac{b_{32}}{b_{33}} x_2 + x_3 = \dfrac{g_3}{b_{33}} \end{cases} \tag{1.45}$$

Equations (1.45) are written in matrix notation, where $\mathbf{X} = \mathbf{AX} + \mathbf{F}$

$$\mathbf{A} = \begin{pmatrix} 0 & -\dfrac{b_{12}}{b_{11}} & - \\[2ex] -\dfrac{b_{21}}{b_{22}} & 0 & - \\[2ex] - & - & 0 \end{pmatrix} \; ; \quad \mathbf{F} = \begin{pmatrix} \dfrac{g}{b_{11}} \\[2ex] \dfrac{g_2}{b_{22}} \\[2ex] \dfrac{g_3}{b_{33}} \end{pmatrix} \tag{1.46}$$

Note the diagonal elements of zero in matrix **A** of (1.46), since equation (1.45) is solved for values of x_1 through x_n which appear along the diagonal. Convergence will be affected with the following norm conditions

$$\sum_{j=1}^{n} \left| \frac{b_{ij}}{b_{ii}} \right| < 1 \qquad i = 1, 2, 3, \ldots, n$$

$$\sum_{i=1}^{n} \left| \frac{b_{ij}}{b_{ii}} \right| < 1 \quad j = 1, 2, 3, \ldots, n \tag{1.47}$$

$$\sum_{i,j=1}^{n} \left| \frac{b_{ij}}{b_{ii}} \right|^2 < 1$$

Note that these norms, much like the vector norms described previously, express the sum or the sum of the squares of the row or column elements of the matrix. Note also that the norms according to the previously given theorem are *sufficient* condition for the convergence. These norms are *not* the necessary condition, which means that a system of equations not satisfying these norms may also converge to a solution.

Gauss-Seidel Iteration Process

Consider the numerical example with two unknowns x_1 and x_2 and two equations

$$3x_1 + x_2 = 7$$
$$x_1 + 3x_2 = 5 \tag{1.48}$$

Using the iterative method of Gauss, we get

$$\begin{cases} x_1 = 2.333 - 0.333x_2 \\ x_2 = 1.666 - 0.333x_1 \end{cases} ; \quad \mathbf{A} = \begin{pmatrix} 0 & -0.333 \\ -0.333 & 0 \end{pmatrix}$$

$$\mathbf{X}^0 = \begin{pmatrix} 2.333 \\ 1.666 \end{pmatrix}; \quad \mathbf{X}^1 = \begin{pmatrix} 1.778 \\ 0.889 \end{pmatrix}; \quad \mathbf{X}^2 = \begin{pmatrix} 2.036 \\ 1.073 \end{pmatrix} \tag{1.49}$$

$$\mathbf{X}^3 = \begin{pmatrix} 1.975 \\ 0.988 \end{pmatrix}; \quad \mathbf{X}^4 = \begin{pmatrix} 2.000 \\ 1.000 \end{pmatrix}$$

Note the difference between the above iteration process and the *Gauss-Siedel* process for the same problem,

$$x_1 = 2.333 - 0.333x_2; \quad x_1 = 2.333, \ x_2 = 0$$
$$x_2 = 1.666 - 0.333x_1; \quad x_2 = 0.889$$

$$\mathbf{X}^0 = \begin{pmatrix} 2.333 \\ 0.889 \end{pmatrix}; \mathbf{X}^1 = \begin{pmatrix} 2.037 \\ 0.988 \end{pmatrix} \tag{1.50}$$

$$\mathbf{X}^2 = \begin{pmatrix} 2.004 \\ 0.999 \end{pmatrix}; \mathbf{X}^3 = \begin{pmatrix} 2.000 \\ 1.000 \end{pmatrix}$$

In the Gauss-Seidel iteration process, the newly calculated values of the components of **X** are used as soon as available, while in the iterative method of Gauss the values of components of **X** are used only after all of the component values are calculated. Note the difference between vectors $\mathbf{X}^2$ in equations (1.49) and (1.50).

Matrix Equation for Gauss-Seidel Process

In the Gauss-Seidel method, the values of the unknowns at iteration $r + 1$ are computed by using existing values from iteration r or iteration $r + 1$, if available. Denoting the iteration numbers by superscripts, we get

$$\begin{aligned} c_{11}x_1^{r+1} + c_{12}x_2^r + c_{13}x_3^r &= b_1 \\ c_{21}x_1^{r+1} + c_{22}x_2^{r+1} + c_{23}x_3^r &= b_2 \\ c_{31}x_1^{r+1} + c_{32}x_2^{r+1} + c_{33}x_3^{r+1} &= b_3 \end{aligned} \tag{1.51}$$

Note that in the first equation we compute x_1 for iteration $r + 1$ and in the second equation we use this value of x_1 and we compute x_2 at iteration $r +$ 1. This is different from Gauss's iteration process, where only the values of x at iteration r are used to compute the values of x at iteration $r + 1$.

We may write equation (1.51) in matrix notation,

$$\mathbf{L}\mathbf{X}^{r+1} + \mathbf{U}\mathbf{X}^r = \mathbf{B} \tag{1.52}$$

where **L** and **U** represent lower and upper triangular matrices. The upper triangular matrix **U** will have diagonal elements of zero. From (1.52), we get

$$\mathbf{X}^{r+1} = \mathbf{L}^{-1}\mathbf{B} - \mathbf{L}^{-1}\mathbf{U}\mathbf{X}^r \tag{1.53}$$

Denoting the product matrices $\mathbf{L}^{-1}\mathbf{B} = \mathbf{A}$ and $\mathbf{L}^{-1}\mathbf{U} = -\mathbf{W}$, we may write

$$\mathbf{X}^{r+1} = \mathbf{A} + \mathbf{W}\mathbf{X}^r \tag{1.54}$$

Using the above relation and denoting the assumed solution (iteration zero) by $\mathbf{X}^0$ and subsequent iterations by $\mathbf{X}^1$, $\mathbf{X}^2$, and so on, we may write,

$$\begin{aligned} \mathbf{X}^1 &= \mathbf{A} + \mathbf{W}\mathbf{X}^0 \\ \mathbf{X}^2 &= \mathbf{A} + \mathbf{W}(\mathbf{A} + \mathbf{W}\mathbf{X}^0) \\ &\vdots \\ \mathbf{X}^r &= (\mathbf{I} + \mathbf{W} + \mathbf{W}^2 + \ldots + \mathbf{W}^{r-1})\mathbf{A} + \mathbf{W}^r\mathbf{X}^0 \end{aligned} \tag{1.55}$$

The goemetric series within the parentheses can be summed for $r \to \infty$ under the condition that the absolute values of the *characteristic values* of **W** be less than unity. Under this condition

$$(\mathbf{I} - \mathbf{W})^{-1} = \mathbf{I} + \mathbf{W} + \mathbf{W}^2 + \mathbf{W}^3 + \ldots \tag{1.56}$$

Using this, the final value of **X** can be calculated as

$$\mathbf{X} = \lim_{r \to \infty} \mathbf{X}^r = (\mathbf{I} - \mathbf{W})^{-1} \mathbf{A} \tag{1.57}$$

The preceding discussion gives rise to the convergence theorem of the Gauss-Seidel process.

Theorem: The Gauss-Seidel process produces solution **X** for the system **CX** = **B** only if each characteristic value of matrix **W** is of modulus (absolute value) of less than unity.

Comparing the above theorem to the one given for the Gauss iteration process, one notes the similarity of these theorems. Observe also that the theorem following equations (1.43) is concerned with the characteristic values of the **A** matrix, while the above theorem deals with the characteristic values of the **W** matrix.

Direct Methods

The direct (noniterative) methods of solution of a set of linear simultaneous equations consist of a set of logical procedures directed toward the eventual solution. The simplest direct method for the solution of linear simultaneous equations is the method of elimination described in the following example. This example consists of two equations and two unknowns,

$$\begin{aligned} 2x_1 + 3x_2 &= 8 \\ -x_1 + x_2 &= 1 \end{aligned} \tag{1.58}$$

The idea of the elimination method is to reduce the square matrix of the coefficients

$$\begin{pmatrix} 2 & 3 \\ -1 & 1 \end{pmatrix} \tag{1.59}$$

to an upper triangular matrix

$$\begin{pmatrix} - & - \\ 0 & - \end{pmatrix} \tag{1.60}$$

where dashes specify numbers. At this point we will have a set of two equations and two unknowns, with the last equation having only one unknown. We solve the last equation for this unknown and proceed to solve the next to the last equation for the next unknown and so on. To carry out this procedure for the case of equation (1.58) we continue as follows.

Divide the first equation by 2 to obtain a unity coefficient for x_1,

$$x_1 + 1.5x_2 = 4$$
$$-x_1 + x_2 = 1 \tag{1.61}$$

Eliminate x_1 from the second equation by adding the first and the second equations.

$$x_1 + 1.5x_2 = 4$$
$$2.5x_2 = 5 \tag{1.62}$$

Now we have reduced the matrix of the coefficients to an upper triangular form.

The last equation results in $x_2 = 2$. Using this in the first equation, we get $x_1 = 1$. The above method can be extended to a set containing any number of linear equations.

Square Root Method

Consider the set of simultaneous equations written in matrix form $\mathbf{AX} = \mathbf{B}$ and assume matrix $\mathbf{A}$ can be written as the product of lower and upper triangular matrices $\mathbf{L}$ and $\mathbf{U}$. In this case, we have

$$\mathbf{AX} = \mathbf{B}; \quad \mathbf{LUX} = \mathbf{B}; \quad \mathbf{X} = \mathbf{U}^{-1}(\mathbf{L}^{-1}\mathbf{B}) \tag{1.63}$$

Choleski's method for matrix triangulation can be used for the solution of sets of linear equations. In this application the method is usually referred to as the square root method.

In addition, consider the case where matrix $\mathbf{A}$ can be written as the product of two transposed triangular matrices $\mathbf{A} = \mathbf{S}'\mathbf{S}$, where prime denotes transpose matrix,

$$\mathbf{S} = \begin{pmatrix} s_{11} & s_{12} & s_{13} \\ 0 & s_{22} & s_{23} \\ 0 & 0 & s_{33} \end{pmatrix} \tag{1.64}$$

From the information given in describing Cholesky's method, we have

$$s_{11} = \sqrt{a_{11}},\ s_{ij} = a_{ij}/s_{11},\ s_{ii} = \sqrt{a_{ii} - \sum_{l=1}^{i=1} s_{li}^2}; \quad i > 1$$

$$s_{ij} = \left(a_{ij} - \sum_{l=1}^{i-1} s_{li}s_{lj}\right) \Big/ s_{ii};\ s_{ij} = 0; \quad i > j \tag{1.65}$$

By defining the column matrix $\mathbf{K} = \mathbf{SX}$, we may write the following additional relationships

$$\begin{aligned} \mathbf{AX} &= \mathbf{B} \\ \mathbf{K} &= \mathbf{SX} \\ \mathbf{S'K} &= \mathbf{B} \end{aligned} \tag{1.66}$$

$$k_1 = \frac{b_1}{s_{11}}\,;\; k_i = \left(b_i - \sum_{l=1}^{i-1} s_{li}k_l\right) \Big/ s_{ii}\,; \quad i > 1$$

where ks are the elements of the $\mathbf{K}$ matrix. Using this, we can write equations for the unknowns $x_1, x_2, \ldots$

$$x_n = \frac{k_n}{s_{nn}}\,;\; x_i = \left(k_1 - \sum_{l=i+1}^{n} s_{il}x_l\right) \Big/ s_{ii}\,; \quad i < n \tag{1.67}$$

The writing of compact equations, such as the ones in (1.65) and (1.66), is elegant and helps the understanding of general principles; in practice, however, one usually starts from basic equations (1.63). Starting from these equations, one may solve a set of three equations in three unknowns and discover the procedure in obtaining the upper and lower triangular matrices, their inverse, and so on. This will lead to following procedures parallel to those of equations (1.65) and (1.66) and will help the understanding of the logic of calculations. This calculation logic is eventually programmed for computer solution.

Example. Consider the set of equations,

$$\begin{aligned} x_1 + x_2 &= 1 \\ x_1 + 2x_2 &= 1 \end{aligned} \tag{1.68}$$

to be solved for x_1 and x_2 using the square root method. In matrix notation we write,

$$\mathbf{A} = \begin{bmatrix} 1 & 1 \\ 1 & 2 \end{bmatrix} = \mathbf{S'S} = \begin{bmatrix} 1 & 0 \\ 1 & 1 \end{bmatrix} \begin{bmatrix} 1 & 1 \\ 0 & 1 \end{bmatrix}$$

$$k_1 = 1, \quad k_2 = b_2 - s_{12}k_1 = 0 \tag{1.69}$$

$$x_2 = 0, \quad x_1 = 1 \text{ or } \begin{pmatrix} x_1 \\ x_2 \end{pmatrix} = \begin{pmatrix} 1 \\ 0 \end{pmatrix}$$

The last matrix on the right gives the values of the unknowns as $x_1 = 1$ and $x_2 = 0$.

Elimination Method and Pivotal Compensation

In the solution of sets of linear simultaneous equations, pivotal compensation is used to reduce the magnitude of coefficients and avoid growth of these coefficients by indiscriminate multiplications. Elimination of variables is accomplished through the selection of the equation having the largest coefficient for the particular variable. Reduction of this coefficient to unity reduces the values of other coefficients within the same equation. The following example will illustrate these statements.

Consider the example of the set of three equations and three unknowns,

$$\begin{aligned} x_1 + 2x_2 + 3x_3 &= 1 \\ 3x_1 + x_2 + x_3 &= 0 \\ 2x_1 + x_2 + x_3 &= 0 \end{aligned} \tag{1.70}$$

The preceding equations are written in matrix form

$$\begin{bmatrix} 1 & 2 & 3 & 1 \\ \underline{3} & 1 & 1 & 0 \\ 2 & 1 & 1 & 0 \end{bmatrix} \tag{1.71}$$

where the right side of (1.70) forms the last column of the matrix. The *pivotal* element, the largest element of the first column, is underlined in equation (1.71).

Exchanging the first and the second rows, we get

$$\begin{bmatrix} \underline{3} & 1 & 1 & 0 \\ 1 & 2 & 3 & 1 \\ 2 & 1 & 1 & 0 \end{bmatrix} \tag{1.72}$$

Dividing the first row by pivotal element and eliminating variable x_1 from second and the third rows, we get

$$\begin{bmatrix} 1 & 1/3 & 1/3 & 0 \\ 0 & \underline{5/3} & 8/3 & 1 \\ 0 & 1/3 & 1/3 & 0 \end{bmatrix} \tag{1.73}$$

Where the second pivotal element 5/3 is underlined. Dividing the second row by 5/3 and eliminating x_2 from the first and the third rows, we get

$$\begin{bmatrix} 1 & 0 & -1/5 & -1/5 \\ 0 & 1 & 8/5 & 3/5 \\ 0 & 0 & -1/5 & -1/5 \end{bmatrix} \tag{1.74}$$

Dividing the third row by $-1/5$ and eliminating x_3 from the second and the first row, we get the answers in the fourth column:

$$\begin{bmatrix} 1 & 0 & 0 & 0 \\ 0 & 1 & 0 & -1 \\ 0 & 0 & 1 & 1 \end{bmatrix} \quad \text{where} \quad \begin{pmatrix} x_1 \\ x_2 \\ x_3 \end{pmatrix} = \begin{pmatrix} 0 \\ -1 \\ 1 \end{pmatrix} \tag{1.75}$$

Ill-Conditioned Matrices

A system of equations with the property that small changes in the coefficients will result in large changes in the solution is called an *ill-conditioned* set of equations. Consider the classical example by T.S. Wilson of the system of four equations in four unknowns:

$$\begin{aligned} 10x_1 + 7x_2 + 8x_3 + 7x_4 &= 32 \\ 7x_1 + 5x_2 + 6x_3 + 5x_4 &= 23 \\ 8x_1 + 6x_2 + 10x_3 + 9x_4 &= 33 \\ 7x_1 + 5x_2 + 9x_3 + 10x_4 &= 31 \end{aligned} \tag{1.76}$$

The matrix of the coefficients $\mathbf{A}$ and also the inverse of this matrix $\mathbf{A}^{-1}$ follow.

$$\mathbf{A} = \begin{pmatrix} 10 & 7 & 8 & 7 \\ 7 & 5 & 6 & 5 \\ 8 & 6 & 10 & 9 \\ 7 & 5 & 9 & 10 \end{pmatrix}$$

and

$$\mathbf{A}^{-1} = \begin{pmatrix} 25 & -41 & 10 & -6 \\ -41 & 68 & -17 & 10 \\ 10 & -17 & 5 & -3 \\ -6 & 10 & -3 & 2 \end{pmatrix} \tag{1.77}$$

The correct solution of the system of equations is

$$x_1 = 1,\ x_2 = 1,\ x_3 = 1, \text{ and } x_4 = 1 \tag{1.78}$$

We note a completely wrong solution

$$(9.2,\ -12.6,\ 4.5,\ -1.1)' \tag{1.79}$$

where prime used to denote transpose matrix, satisfies equations (1.76) within ± 0.1. And another set of wrong solutions, completely different from the first set,

$$(6,\ -7.2,\ 2.9,\ -0.1)' \tag{1.80}$$

will result in the right side values of

$$(32.1,\ 22.9,\ 32.9, \text{ and } 31.1)' \tag{1.81}$$

Again these values are within ± 0.1 of the values of equation (1.76).

Thus, we refer to matrix **A** as an ill-conditioned matrix. A measure of numerical difficulties associated with certain matrices is given by the so-called *conditioning numbers*

$$\begin{aligned} M &= n[\max |a_{ij}|][\max |a_{ij}|^{-1}] \\ N &= n^{-1}\left[\sum_{i,j} |a_{ij}|^2\right]^{1/2} \left[\sum_{i,j} (|a_{ij}|^{-1})^2\right]^{1/2} \\ P &= \lambda/\mu \end{aligned} \tag{1.82}$$

Where $|a_{ij}|$ refers to the elements of matrix **A** and $|a_{ij}|^{-1}$ refers to the elements of matrix $\mathbf{A}^{-1}$. The symbol n is the number of simultaneous equations and λ and μ are the largest and smallest characteristic values of matrix **A**.

The conditioning numbers of (1.82) are usually difficult to calculate, especially the characteristic values. The values of equation (1.82) are related to the previously given norms for vectors and matrices.

For matrix **A** of equation (1.77), we get

$$\begin{aligned} M &= 4 \cdot 10 \cdot 68 = 2720 \\ N &= \frac{1}{4}(933)^{1/2}(9708)^{1/2} \\ &= 752 \\ P &= 2984 \end{aligned} \tag{1.83}$$

Usually large conditioning numbers indicate numerical difficulties.

Such difficulties could have been anticipated from the application of the Gauss-Seidel iteration process. Consider the following solution obtained for the set of equations (1.76) using the Gauss-Seidel process:

$$\begin{aligned} \text{Start} \quad & \mathbf{X} = (0, 0, 0, 0)' \\ \text{Iteration (1)} \quad & (3.20, 0.12, 0.67, 0.20)' \\ (2) \quad & (2.44, 0.18, 1.06, 0.35)' \\ (3) \quad & (1.98, 0.21, 1.28, 0.46)' \\ (4) \quad & (1.71, 0.21, 1.39, 0.54)' \\ & \vdots \\ (30) \quad & (1.53, 0.13, 1.22, 0.87)' \end{aligned} \tag{1.84}$$

Even after thirty iterations, no convergence is in sight!

The correct answers, however, could have been readily obtained by using the elimination method or by using

$$\mathbf{X} = \mathbf{A}^{-1}\mathbf{B} \tag{1.85}$$

with the desired result (1, 1, 1, 1).

A somewhat quicker, although not mathematically rigorous, method of predicting convergence by the Gauss-Seidel iteration process is to see if the diagonal elements of matrix **A** dominate; that is, if these elements are larger than the nondiagonal elements.

As a counter example to the above iteration process, consider the previously given set of equations:

$$\begin{aligned} 3x_1 + x_2 &= 7 \\ x_1 + 3x_2 &= 5 \end{aligned} \tag{1.86}$$

which converged to a solution by the fourth iteration. In this case

$$\mathbf{A} = \begin{bmatrix} 3 & 1 \\ 1 & 3 \end{bmatrix}, \qquad \mathbf{A}^{-1} = \begin{bmatrix} 3/8 & -1/8 \\ -1/8 & 3/8 \end{bmatrix}$$

$$|\mathbf{A} - \lambda\mathbf{I}| = \begin{vmatrix} 3 - \lambda & 1 \\ 1 & 3 - \lambda \end{vmatrix} = 0 \quad \text{or} \quad \begin{gathered} \lambda^2 - 6\lambda + 8 = 0 \\ \text{with } \lambda = 4, 2 \end{gathered} \tag{1.87}$$

The conditioning numbers become

$$\begin{aligned} M &= 2(3)(3/8) = 2.25 \\ N &= 1/2(20)^{1/2}(0.3125)^{1/2} = 1.250 \\ P &= 4/2 = 2 \end{aligned} \tag{1.88}$$

Small numbers compared to the previously given case, equations (1.83), of the ill-conditioned matrix!

Inversion of Matrices by Elimination Method

The method of elimination, as given previously for the solution of sets of linear equations, can be expanded and used for matrix inversion. The elimination method of matrix inversion, described later, has the additional advantage of compacting the calculations to utilize a smaller storage capacity. Storage capacity or computer memory considerations are important in numerical calculations.

In general to invert an $n \times n$ matrix **A** one needs n^2 memory locations for the matrix elements of **A** and also n^2 locations for the elements of the inverse matrix $\mathbf{A}^{-1}$, for a total of $2n^2$ locations. The method described takes advantage of arranging the intermediate results in order to reduce the number of required memory locations to $n^2 + n$.

Assume it is required to obtain the inverse of matrix **A** designated by matrix **X**, so that

$$\mathbf{AX} = \mathbf{I} \tag{1.89}$$

where **I** is the unit matrix. Multiplying both sides of (1.89) by $\mathbf{A}^{-1}$, we get

$$\mathbf{X} = \mathbf{A}^{-1}\mathbf{I} \tag{1.90}$$

To obtain $\mathbf{A}^{-1}$ we may use the following procedure, which also uses pivotal compensation.

1. Construct an n by $(n + 1)$ matrix $[\mathbf{A}, 0]$, where the "zero column" is added to the **A** matrix. (See example).
2. Scan the elements of the first column and locate the term having the largest absolute value (*pivotal element*). Designate the corresponding row number by k_1. Replace the zero in the k_1 row of the last column by one.
3. Normalize the *pivotal row* by dividing through by the pivotal element.
4. Using the normalized pivotal row, delete the elements of the first column.
5. Take out the first column and repeat steps 1 through 4.
6. Unscramble the resulting final matrix according to the following rules:

$$\begin{pmatrix} 1 & 2 & 3 & \dots & n \\ \\ k_1 & k_2 & k_3 & \dots & k_n \end{pmatrix} \tag{1.91}$$

$$\begin{array}{lcl} \text{Col } 1 \to \text{col } k_1 & & \text{row } k_1 \to \text{row } 1 \\ \vdots & \text{and} & \vdots \\ \text{Col } n \to \text{col } k_n & & \text{row } k_n \to \text{row } n \end{array}$$

Example: Find the inverse of the matrix

$$\begin{bmatrix} 1 & 10 & 1 \\ 2 & 0 & 1 \\ 3 & 3 & 2 \end{bmatrix} \tag{1.92}$$

Form the matrix of step 1 and locate the pivotal element (underlined).

$$\begin{bmatrix} 1 & 10 & 1 & 0 \\ 2 & 0 & 1 & 0 \\ \underline{3} & 3 & 2 & 1 \end{bmatrix}; \quad k_1 = 3 \tag{1.93}$$

Divide the third row by the pivotal element and eliminate the corresponding elements from rows 1 and 2.

$$\begin{bmatrix} 0 & \underline{9} & 1/3 & -1/3 \\ 0 & -2 & -1/3 & -2/3 \\ 1 & 1 & 2/3 & 1/3 \end{bmatrix}; \quad k_2 = 1 \tag{1.94}$$

Take the second through fourth columns of the matrix above and perform steps 1 and 2.

$$\begin{bmatrix} \underline{9} & 1/3 & -1/3 & 1 \\ -2 & -1/3 & -2/3 & 0 \\ 1 & 2/3 & 1/3 & 0 \end{bmatrix} \tag{1.95}$$

Divide the first row by the pivotal element and eliminate corresponding elements from the second and the third rows. This step and the rest of the procedure follows.

$$\begin{bmatrix} 1 & 1/27 & -1/27 & 1/9 \\ 0 & \underline{-7/27} & -20/27 & 2/9 \\ 0 & 17/27 & 10/27 & -1/9 \end{bmatrix} k_3 = 2 \tag{1.96}$$

$$\begin{bmatrix} 1/27 & -1/27 & 1/9 & 0 \\ \underline{-7/27} & -20/27 & 2/9 & 1 \\ 17/27 & 10/27 & -1/9 & 0 \end{bmatrix} \tag{1.97}$$

$$\begin{bmatrix} 0 & -1/7 & 1/7 & 1/7 \\ 1 & 20/7 & -6/7 & -27/7 \\ 0 & -10/7 & 3/7 & 17/7 \end{bmatrix} \tag{1.98}$$

With the answers, prior to permutation, given in rows 2, 3, and 4 of (1.98),

$$\begin{bmatrix} -1/7 & 1/7 & 1/7 \\ 20/7 & -6/7 & -27/7 \\ -10/7 & 3/7 & 17/7 \end{bmatrix} \tag{1.99}$$

Using the permutation schedule

$$\begin{pmatrix} 1 & 2 & 3 \\ 3 & 1 & 2 \end{pmatrix} \tag{1. 100}$$

we get the inverse

$$\mathbf{A}^{-1} = \begin{bmatrix} 3/7 & 17/7 & -10/7 \\ 1/7 & 1/7 & -1/7 \\ -6/7 & -27/7 & 20/7 \end{bmatrix} \tag{1.101}$$

Another method of solving simultaneous equations is to obtain the inverse of the coefficients matrix and multiply their inverse by the matrix of the right side, as given in equation (1.85).

Correction of the Elements of the Inverse Matrix

Assume $\mathbf{D}_0$ is calculated and is the matrix representing $\mathbf{A}^{-1}$. Then $\mathbf{I} - \mathbf{AD}_0$ will be the error matrix $\mathbf{F}_0$. In this iterative method, $\mathbf{D}_0$ can also be an approximate inverse. Consider the following relations

$$\mathbf{F}_0 = \mathbf{I} - \mathbf{AD}_0 \tag{1.102}$$

For the first iteration we compute

$$\mathbf{D}_1 = \mathbf{D}_0(\mathbf{I} + \mathbf{F}_0)\,; \quad \mathbf{F}_1 = \mathbf{I} - \mathbf{AD}_1 \tag{1.103}$$

and continue with the second, third, . . . iterations,

$$\mathbf{D}_2 = \mathbf{D}_1(\mathbf{I} + \mathbf{F}_1)\,; \quad \mathbf{F}_2 = \mathbf{I} - \mathbf{AD}_2$$

$$\vdots$$

$$\mathbf{D}_m = \mathbf{D}_{m-1}(\mathbf{I} + \mathbf{F}_{m-1})\,; \quad \mathbf{F}_m = \mathbf{I} - \mathbf{AD}_m \tag{1.104}$$

We also have

$$\mathbf{F}_m = \mathbf{I} - \mathbf{AD}_m = \mathbf{I} - \mathbf{AD}_{m-1}(\mathbf{I} + \mathbf{F}_{m-1}) \tag{1.105}$$

Noting that $\mathbf{F}_{m-1} = \mathbf{I} - \mathbf{AD}_{m-1}$ and from this $\mathbf{AD}_{m-1} = \mathbf{I} - \mathbf{F}_{m-1}$, and replacing this in (1.105), we get

$$\begin{aligned} \mathbf{F}_m &= \mathbf{I} - (\mathbf{I} - \mathbf{F}_{m-1})(\mathbf{I} + \mathbf{F}_{m-1}) \\ &= \mathbf{I} - (\mathbf{I}^2 - \mathbf{F}_{m-1}^2) \\ &= \mathbf{F}_{m-1}^2 \end{aligned} \tag{1.106}$$

or, generalizing,

$$\mathbf{F}_m = \mathbf{F}_{m-1}^2 = \mathbf{F}_{m-2}^4 = \ldots = \mathbf{F}_0^{2m} \tag{1.107}$$

Thus, the last estimate of the inverse matrix $\mathbf{D}_m$ becomes

$$\mathbf{D}_m = \mathbf{A}^{-1}(\mathbf{I} - \mathbf{F}_0^{2m}) \tag{1.108}$$

Assuming the norms of $\mathbf{F}_0$ are such as to make its powers convergent, the matrix $\mathbf{D}_m$ will approach the inverse matrix $\mathbf{A}^{-1}$ with a convergence rate proportional to the square of the initial error matrix $\mathbf{F}_0$.

Example: As an example of the proceeding, consider the matrix

$$\mathbf{A} = \begin{pmatrix} 5 & 2 \\ 3 & -1 \end{pmatrix} \tag{1.109}$$

We have obtained an approximate inverse

$$\mathbf{D}_0 = \begin{pmatrix} 0.1 & 0.2 \\ 0.3 & -0.4 \end{pmatrix} \tag{1.110}$$

Following these procedures, we have

$$\mathbf{F}_0 = \begin{pmatrix} 1 & 0 \\ 0 & 1 \end{pmatrix} - \begin{pmatrix} 5 & 2 \\ 3 & -1 \end{pmatrix} \begin{pmatrix} 0.1 & 0.2 \\ 0.3 & -0.4 \end{pmatrix}$$

$$= \begin{pmatrix} -0.1 & -0.2 \\ 0 & 0 \end{pmatrix} \tag{1.111}$$

$$\mathbf{D}_1 = \begin{pmatrix} 0.1 & 0.2 \\ 0.3 & -0.4 \end{pmatrix} \begin{pmatrix} 0.9 & -0.2 \\ 0 & 1 \end{pmatrix} = \begin{pmatrix} 0.09 & 0.18 \\ 0.27 & -0.46 \end{pmatrix} \tag{1.112}$$

Similarly

$$\mathbf{D}_2 = \begin{pmatrix} 0.091 & 0.182 \\ 0.273 & -0.454 \end{pmatrix} \tag{1.113}$$

This procedure can be carried out until desired accuracy is reached.

Evaluation of the Methods

The Gauss-Seidel method of iteration requires n^2 multiplications and/or divisions per iteration for a set of n equations and n unknowns. The number of iterations, of course, depends on the conditioning matrix and its norms. The method of elimination requires $n^3/3$ multiplications and/or divisions for the solution of a set of n equations. Matrix inversion usually requires in the order n^3 multiplications and/or divisions.

Given a set of simultaneous equations, one is usually faced with deciding upon the most efficient method of solution. A variety of methods is given in this chapter, each requiring a varying amount of logical operations for their implementation. The convergence criteria of some of these methods are not, usually, readily implemented. In engineering applications, one usually starts by applying the simplest method of solution. If that doesn't work, one may proceed to more sophisticated methods.

The direct method of solution given by the example of equation (1.58) is most easily implemented through a digital computer program. If coefficients of the set of simultaneous equations are such that calculation errors resulting in unacceptable answers are encountered, one may use the pivotal compensation method of equation (1.70). Note that the logic of programming the pivotal compensation method is considerably more complicated than the direct method. Iteration methods, because of uncertainty of the number of iterations required, are usually used as a last resort. For this reason, and also because of the fact that their convergence criteria are not easily implemented, the use of iteration methods in the solution of simultaneous equations is not widespread.

The inversion of matrices, as seen from the discussion in this chapter, is closely related to the solution of sets of simultaneous equations. Elimination method, with or without pivotal compensation, is usually used for matrix inversion. The method described for the correction of the elements of inverse matrix is also implemented in several digital computer installations.

Problems

1. Given the 3×3 matrix

$$\begin{pmatrix} 10 & 16 & -2 \\ 12 & 8 & 24 \\ 13 & -2 & 6 \end{pmatrix} = \mathbf{LU}$$

obtain the lower **L** and upper **U** triangular matrices. Set the diagonal elements of **L** equal to unity.

2. Given the 3 by 3 matrix

$$\begin{pmatrix} 16 & 5 & -3 \\ 6 & 8 & 12 \\ 12 & -2 & 18 \end{pmatrix} = \mathbf{LU}$$

obtain the lower **L** and upper **U** triangular matrices. Set the diagonal elements of **L** equal to unity.

3. Find the inverse of matrix

$$\begin{pmatrix} 16 & 8 \\ 5 & -12 \end{pmatrix}$$

by obtaining the lower **L** and upper **U** triangular matrices (set the diagonal elements of **L** equal to unity) and their inverse products.

4. Find the inverse of matrix

$$\begin{pmatrix} 8 & 12 \\ 86 & -3 \end{pmatrix}$$

by finding the lower **L** and upper **U** triangular matrices (set the diagonal elements of **L** equal to unity) and their inverse products.

5. Find the inverse of the symmetrical matrix

$$\begin{pmatrix} 2 & 8 \\ 8 & -3 \end{pmatrix}$$

using Choleski's method.

6. Find the inverse of the symmetrical matrix

$$\begin{pmatrix} 2 & 6 & 8 \\ 6 & -3 & 12 \\ 8 & 12 & 6 \end{pmatrix}$$

using Choleski's method.

7. Find the inverse of the symmetrical matrix

$$\begin{pmatrix} 12 & -3 & 4 \\ -3 & 8 & 7 \\ 4 & 7 & 6 \end{pmatrix}$$

using Choleski's method.

8. Using the iteration method of Gauss, obtain the first three iterations for the set of equations

$$6x_1 - x_2 = 48$$

$$x_2 + 12x_2 = 154$$

9. Using the iteration method of Gauss, obtain the first three iterations for the set of equations

$$3x_1 - x_2 = 18$$

$$x_1 + 18x_2 = 57$$

10. Using the iteration method of Gauss, obtain the first three iterations for the set of equations

$$6x_1 + 7x_2 + 3x_3 = 29$$

$$x_1 - 3x_2 + 7x_3 = 16$$

$$2x_1 + 8x_2 - 3x_3 = 9$$

11. Using the Gauss-Seidel iteration process, find the first three iterations for the set of equations

$$6x_1 - x_2 = 48$$

$$x_1 + 12x_2 = 154$$

12. Using the Gauss-Seidel iteration process, find the first three iterations for the set of equations

$$3x_1 - x_2 = 18$$

$$x_1 + 18x_2 = 57$$

13. Using the Gauss-Seidel iteration process, find the first three iterations for the set of equations

$$6x_1 + 7x_2 + 3x_3 = 29$$

$$x_1 - 3x_2 + 7x_3 = 16$$

$$2x_1 + 8x_2 - 3x_3 = 9$$

14. Using the Gauss-Seidel iteration process, find the first three iterations for the set of equations

$$8x_1 + 7x_2 + 13x_3 = 48$$

$$x_1 - 13x_2 + 16x_3 = 53$$

$$-8x_1 + 6x_2 + 23x_3 = -13$$

15. Solve the set of simultaneous equations

$$3x_1 - 6x_2 + 7x_3 = -35$$

$$x_1 - 3x_2 + x_3 = -12$$

$$2x_1 + 3x_2 - 6x_3 = 19$$

by the direct method of solution.

16. Solve the set of simultaneous equations

$$\begin{aligned} x_1 - 2x_2 + x_3 &= -1 \\ x_1 + 3x_2 - 7x_3 &= 22 \\ 2x_1 - 7x_2 + 4x_3 &= -13 \end{aligned}$$

by the direct method of solution.

17. Solve the set of equations

$$\begin{aligned} -3x_1 + 7x_2 &= -10 \\ 7x_1 + 2x_2 &= +5 \end{aligned}$$

by the square root method.

18. Solve the set of equations

$$\begin{aligned} 3x_1 + x_2 &= 12 \\ x_2 + 8x_1 &= 22 \end{aligned}$$

by the square root method.

19. Using the elimination method and pivotal compensation, solve the set of equations of problem 13.
20. Using the elimination method and pivotal compensation, solve the set of equations of problem 14.
21. Using the elimination method and pivotal compensation, solve the set of equations of problem 15.
22. Invert the matrix

$$\begin{pmatrix} 16 & 7 \\ 2 & -3 \end{pmatrix}.$$

by the elimination method.

23. Invert the matrix

$$\begin{pmatrix} 1 & -2 & 1 \\ 1 & 3 & -7 \\ 2 & -7 & 4 \end{pmatrix}$$

by the elimination method.

24. Invert the matrix

$$\begin{pmatrix} 6 & 7 & 3 \\ 1 & -3 & 7 \\ 2 & 8 & -3 \end{pmatrix}$$

by the elimination method.

25. The approximate inverse of the matrix

$$\mathbf{A} = \begin{bmatrix} 1 & 1 & 1 \\ 1 & 2 & 2 \\ 1 & 2 & 3 \end{bmatrix}$$

has been found to be

$$\mathbf{A}^{-1} = \begin{bmatrix} 2.1 & -0.9 & 0 \\ -1.1 & 2.0 & -1.0 \\ 0 & -1.0 & 1.1 \end{bmatrix}$$

Improve this inverse by two correction iterations.

26. The approximate inverse of the matrix

$$\mathbf{A} = \begin{bmatrix} 2 & 1 & 1 \\ 1 & 3 & 1 \\ 1 & 1 & 4 \end{bmatrix}$$

has been found to be

$$\mathbf{A}^{-1} = \begin{bmatrix} 0.60 & -0.18 & -0.12 \\ -0.18 & 0.41 & -0.06 \\ -0.12 & -0.06 & 0.29 \end{bmatrix}$$

Obtain the improved value of the inverse after two corrective iterations.

27. Find the norms given in equation (1.82) for the matrix

$$\begin{pmatrix} 27 & 33.2 \\ 17.6 & 3.7 \end{pmatrix}$$

28. Find the norms given in equation (1.82) for the matrix

$$\begin{pmatrix} 21.3 & 17.6 \\ 1.8 & 19.7 \end{pmatrix}$$

Bibliography

Fröberg, C.W. *Numerical Analysis*. Reading, Mass.: Addison Wesley, 1965.

Hovanessian, S.A., and Pipes, L.A. *Digital Computer Methods*. New York: McGraw-Hill, 1969.

Pipes, L.A., and Hovanessian, S.A. *Matrix Computer Methods in Engineering*, New York: Wiley, 1969.

Rosen, D.J., and Willoughby, R.A. *Sparse Matrices*, New York: Plenum Press, 1972.

Rust, B.M., and Burrus, W.R. *Mathematical Programming and Numerical Solution of Linear Equations*. New York: Elsevier, 1972.

Stewart, J.W., *Introduction to Matrix Computations*, New York: Academic Press, 1973.

Steinberg, D.I. *Computational Matrix Algebra*. New York: McGraw-Hill, 1974.

2 Calculus of Finite Differences

In the numerical solution of mathematical problems, calculations are performed at discrete time or space intervals. The numerical values of the solutions are obtained at these intervals. This is to be distinguished from classical solutions where a continuous solution in the form of an equation is derived and is evaluated for the given values of the independent variable. To illustrate this, consider the second-order linear differential equation

$$\frac{d^2y}{dt^2} + y = 0 \tag{2.1}$$

where t is time and the initial conditions are $y = 1$, $dy/dt = 0$ at $t = 0$. The solution of this equation is of the form given in Boyce and DiPrima.[1]

$$y = A \sin t + B \cos t \tag{2.2}$$

where A and B are constants, to be evaluated with initial conditions.

The solution can also be obtained intuitively, using the following reasoning. Equation (2.1) required that the sum of the second derivative of a function and the function itself be equal to zero. Functions satisfying this condition are trigonometric functions *sine* and *cosine*. In addition, since equation (2.1) is a differential equation of the second order, it should contain *two* constants in the solution. This condition will require both sine and cosine to be present in the solution as given above.

Setting the initial condition $dy/dt = 0$ at $t = 0$, we get

$$\begin{aligned} \frac{dy}{dt} &= A \cos t - B \sin t\Big]_{t=0} \\ &= A \\ &= 0 \end{aligned} \tag{2.3}$$

Subsequently, setting the condition $y = 1$ at $t = 0$ we get

$$\begin{aligned} y &= B \cos t\Big]_{t=0} \\ &= B \\ &= 1 \end{aligned} \tag{2.4}$$

Or the final solution becomes,

$$y = \cos t \tag{2.5}$$

This solution gives the value of y for any value of time t. On the other hand, the numerical solution of differential equation (2.1) is obtained by selecting a time interval Δt, transforming the equation into difference form (which will be discussed later) and obtaining a recursive equation for y in terms of two values of y at preceding time intervals. Or, in functional form

$$y_{i+1} = f(y_i, y_{i-1}, \Delta t) \tag{2.6}$$

the indexes i and $i - 1$ refer to values of y at time t_i and $t_{i-1}(t_i - t_{i-1} = \Delta t)$ currently available and $i + 1$ refers to the value of y at t_{i+1} to be obtained. Thus, the application of the above relation together with the initial conditions will result in values of y at 0, Δt, $2\Delta t$, $3\Delta t$, . . . and so on. It is important to note that, unlike solution (2.5) the value of y at $t = n\Delta t$ (where n is an integer) cannot be obtained unless the previous values of y are available.

It is seen that the numerical calculations involved in obtaining the value of y from (2.6) at a given time t are considerably more than the evaluation of solution (2.5). This is true in the example, since an analytical closed-form solution exists for the differential equation (2.1). However, many engineering applications of numerical methods involve nonlinear differential equations, polynomial interpolation of tabulated functions, and the solution of transcendental equations where no analytical closed-form solutions exist. In these cases, solutions are obtained by numerical formulation of the problems and application of digital computers for evaluation of solutions.

It should also be pointed out that in most engineering problems, even if closed-form solutions are available, they should be evaluated and plotted, while the numerical methods result in solutions already in tabulated or graphical forms.

Difference Operators

In numerical solutions, the *difference* operators are used in a manner similar to that of *differential* or Laplace operators (usually written as D or S) in the differential equations. A complete familiarity with the difference operator is considered essential for an understanding of the underlying principles of numerical methods.

To define various difference operators, consider the function $y = f(x)$

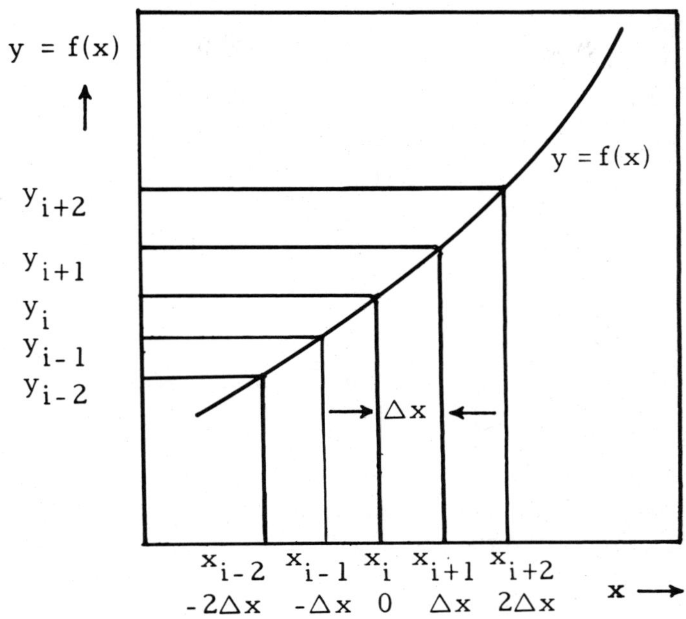

Figure 2-1. The Function $y = f(x)$

described in Figure 2-1 with equal intervals Δx for the independent variable x.

The first *backward difference* of y_i is defined as

$$\nabla y_i = y_i - y_{i-1} \tag{2.7}$$

where ∇ denotes the backward difference operator.

The second backward difference can be obtained from the first:

$$\begin{aligned} \nabla^2 y_i &= \nabla(\nabla y_i) = \nabla(y_i - y_{i-1}) \\ &= \nabla y_i - \nabla y_{i-1} \\ &= (y_i - y_{i-1}) - (y_{i-1} - y_{i-2}) \\ &= y_i - 2y_{i-1} + y_{i-2} \end{aligned} \tag{2.8}$$

Similarly, the third backward difference becomes

$$\nabla^3 y_i = y_i - 3y_{i-1} + 3y_{i-2} - y_{i-3} \tag{2.9}$$

The first *forward difference* of y_i is defined as

$$\Delta y_i = y_{i+1} - y_i \tag{2.10}$$

where Δ denotes the forward difference operator. The second and third forward difference operators can be obtained from the first and second, respectively:

$$\Delta^2 y_i = \Delta(\Delta y_i)$$
$$= \Delta y_{i+1} - \Delta y_i$$
$$= y_{i+2} - 2y_{i+1} + y_i \tag{2.11}$$
$$\Delta^3 y_i = \Delta(\Delta^2 y_i)$$
$$= y_{i+3} - 3y_{i+2} + 3y_{i+1} - y_i$$

The first *central difference* operator of y_i is defined as

$$\delta y_i = y_{i+1/2} - y_{i-1/2} \tag{2.12}$$

where δ denotes the central difference operator. Note that the first central difference involves the values of the function at half intervals. The second central difference operator is obtained from the first:

$$\delta^2 y_i = \delta(\delta y_i) = \delta(y_{i+1/2} - y_{i-1/2})$$
$$= (y_{i+1/2+1/2} - y_{i+1/2-1/2}) - (y_{i-1/2+1/2} - y_{i-1/2-1/2})$$
$$= y_{i+1} - 2y_i + y_{i-1} \tag{2.13}$$

Similarly, third and fourth central difference operators can be obtained:

$$\delta^3 y_i = \delta(\delta^2 y_i) = y_{i+3/2} - 3y_{i+1/2} + 3y_{i-1/2} - y_{i-3/2}$$
$$\delta^4 y_i = \delta(\delta^3 y_i) = y_{i+2} - 4y_{i+1} + 6y_i - 4y_{i-1} + y_{i-2} \tag{2.14}$$

As a numerical example of the above equations, consider the calculation of difference operators of the equation $y = x^2$ as given in Table 2-1. In this table, the second and higher differences are obtained by arithmetical operations on the lower differences; for example $\nabla^2 y = \nabla(\nabla y)$ or $\nabla^2 y = 5 - 3 = 2$. Only the value of $\delta^2 y$ is given, since δy will required values of y at half interval points ($i \pm 1/2$).

Relationship Between Difference and Differential Operators

Let us define the symbolic *difference operator E* as follows:

$$E y_i = y_{i+1}$$
$$E^{-1} y_i = y_{i-1} \tag{2.15}$$
$$E^{\pm 1/2} y_i = y_{i\pm 1/2}$$

Table 2-1
Difference Operators of $y = x^2$

i	x_i	y_i	∇y_i	$\nabla^2 y_i$	$\nabla^3 y_i$	Δy_i	$\Delta^2 y_i$	$\Delta^3 y_i$	$\delta^2 y_i$
1	0					1	2	0	
2	1	1	1			3	2	0	2
3	2	4	3	2		5	2	0	2
4	3	9	5	2	0	7	2	0	2
5	4	16	7	2	0	9	2	0	2
6	5	25	9	2	0	11	2	0	2
7	6	36	11	2	0	13	2	0	2
8	7	49	13	2	0	15	2	0	2
9	8	64	15	2	0	17	2		2
10	9	81	17	2	0	19			2
11	10	100	19	2	0				

Using these relations, we can obtain difference operators ∇, Δ, and δ in terms of E

$$\begin{aligned}
\nabla y_i &= y_i - y_{i-1} = y_i - E^{-1}y_i = \left(1 - \frac{1}{E}\right)y_i \\
\Delta y_i &= y_{i+1} - y_i = Ey_i - y_i = (E - 1)y_i \qquad (2.16) \\
\delta y_i &= y_{i+1/2} - y_{i-1/2} = E^{1/2}y_i - E^{-1/2}y_i = (E^{1/2} - E^{-1/2})y_i
\end{aligned}$$

Equating the operand of y_i in the above equations, we get

$$\begin{aligned}
\nabla &= 1 - \frac{1}{E} \quad \text{backward difference operator} \\
\Delta &= E - 1 \quad \text{forward difference operator} \qquad (2.17) \\
\delta &= E^{1/2} - E^{-1/2} \quad \text{central difference operator}
\end{aligned}$$

The following algebraic relations also hold between the difference operators:

$$\left.\begin{aligned}
\Delta k &= k\Delta \\
Ek &= kE \\
E\Delta &= \Delta E \\
E^n\Delta &= \Delta E^n
\end{aligned}\right\} \text{same for } \nabla \text{ and } \delta \text{ operators} \qquad (2.18)$$

where k and n are constants.

The relation between *difference* operators and the *differential* operator is obtained by writing Taylor's series expansion of the function $f(x + h)$ about x.

$$f(x + h) = f(x) + hf'(x) + \frac{h^2}{2!} f''(x) + \ldots + \frac{h^n}{n!} f^n(x) \tag{2.19}$$

where primes denote derivatives with respect to x. Denoting the *differential operator* $D = d/dx$ and using the *difference operator* E, equation (2.19) becomes

$$Ef(x) = \left[1 + hD + \frac{h^2}{2!} D^2 + \frac{h^3}{3!} D^3 + \ldots\right] f(x) \tag{2.20}$$

Equating the operand of $f(x)$ in the above equation and noting that the infinite series in the brackets represents the series for the exponential function e^{hD}, we get

$$E = e^{hD} \tag{2.21}$$

Taking the natural logarithm of both sides,

$$D = \frac{\ln E}{h} \tag{2.22}$$

These formulas can be applied to obtain relationships between difference and differential operators. These relationships are used in the numerical solution of differential equations.

Relationship Between ∇ and D

The backward difference operator ∇ can be obtained in terms of the differential operator D with the aid of equation (2.17)

$$\nabla = 1 - \frac{1}{E} = 1 - e^{-hD} \tag{2.23}$$

Taylor's series expansion of e^{-hD} will result.

$$\nabla = 1 - \left[1 - hD + \frac{h^2}{2!} D^2 - \frac{h^3}{3!} D^3 + \ldots\right]$$

$$= hD - \frac{h^2}{2!} D^2 + \frac{h^3}{3!} D^3 - \ldots \tag{2.24}$$

Similar expressions can be obtained for higher-order difference operators:

$$\nabla^2 = \left(1 - e^{-hD}\right)^2 = 1 + e^{-2hD} - 2e^{-hD}$$

$$= 1 + \left(1 - \frac{2hD}{1!} + \frac{4h^2D^2}{2!} - \ldots\right)$$

$$-2\left(1 - \frac{hD}{1!} + \frac{h^2D^2}{2!} - \frac{h^3D^3}{3!} + \ldots\right)$$

$$= h^2D^2 - h^3D^3 + \frac{7}{12}h^4D^4 - \ldots \tag{2.25}$$

The product of equations (2.24) and (2.25) will give the value of ∇^3 in terms of differential operator D.

Conversely, the values of differential operator D can be obtained in terms of the difference operator ∇ by using the relation

$$e^{-hD} = 1 - \nabla$$

$$\ln e^{-hD} = -hD = \ln(1 - \nabla) \tag{2.26}$$

Writing the series for $\ln(1 - \nabla)$, we have

$$D = \frac{1}{h}\left(\nabla + \frac{\nabla^2}{2} + \frac{\nabla^3}{3} + \frac{\nabla^4}{4} + \ldots\right) \tag{2.27}$$

Higher-order derivative operators D^2, D^3, ..., can be obtained successively by forming products of lower-order derivatives.

The relationships between ∇ and D operators are summarized:

$$\nabla = hD - \frac{h^2}{2!}D^2 + \frac{h^3}{3!}D^3 - \ldots \tag{2.28}$$

$$\nabla^2 = h^2D^2 - h^3D^3 + \frac{7}{12}h^4D^4 - \ldots \tag{2.29}$$

$$\nabla^3 = h^3D^3 - \frac{3}{2}h^4D^4 + \frac{5}{4}h^5D^5 - \ldots \tag{2.30}$$

$$D = \frac{1}{h}\left(\nabla + \frac{\nabla^2}{2} + \frac{\nabla^3}{3} + \frac{\nabla^4}{4} + \ldots\right) \tag{2.31}$$

$$D^2 = \frac{1}{h^2}\left(\nabla^2 + \nabla^3 + \frac{11}{12}\nabla^4 + \frac{5}{6}\nabla^5 + \ldots\right) \tag{2.32}$$

$$D^3 = \frac{1}{h^3}\left(\nabla^3 + \frac{3}{2}\nabla^4 + \frac{7}{4}\nabla^5 + \ldots\right) \tag{2.33}$$

$$D^4 = \frac{1}{h^4}\left(\nabla^4 + 2\nabla^5 + \frac{17}{6}\nabla^6 + \ldots\right) \tag{2.34}$$

These relations give the infinite series representations of backward differences in terms of differentials and vice versa. In using these equations in actual computation, however, we can take only a finite number of terms. Assuming that the series are convergent and that each term is smaller in magnitude than the preceding term (this is usually the case, since the increment h is small and its powers increase), the first term dropped will represent the largest term neglected in the succeeding calculations. This term will give the order of magnitude of terms neglected in the finite difference representation. It should be noted that in truncating an infinite series the sum of all of the terms dropped gives the actual truncation error. Since getting an expression for this sum is a formidable task in many cases, the first term dropped from the series is taken as the order of magnitude of the error introduced. For example, taking the first two terms of equation (2.29) and neglecting the terms $7/12h^4D^4$ and beyond will result in an error equivalent to the absolute value of this term. In addition, since in many problems the value of the derivative of the variable D^4y is also an unknown, this absolute value cannot be computed. For this reason, we make another simplification in describing the error by stating that neglecting the third term and beyond of the series of equation (2.29) introduces an "error of order h^4" in the resulting expression for ∇^2. The error of order h^4 is written symbolically as the error of $O(h^4)$.

From equations (2.28 through 2.34), the derivatives of various functions can be obtained in terms of the available values of the function. Suppose that we want to obtain the value of D in terms of ∇s with error of order h. From equation (2.28), ∇^2 involves terms of h^2 and higher and, since in (2.31) this term, ∇^2, is divided by h, neglecting it in (2.31) will result in an error of order h, as follows:

$$D = \frac{\nabla}{h}, \quad Dy_i = \frac{1}{h}\nabla y_i$$

$$= \frac{y_i - y_{i-1}}{h}; \quad e = O(h) \tag{2.35}$$

where $e = O(h)$ refers to the error of order h. If the error is to be of order $h^2[e = O(h^2)]$ the ∇^2 term in equation (2.31) should be retained and the ∇^3, ∇^4, ... terms dropped. This, using definitions of ∇ and ∇^2, results in

$$Dy_i = \frac{1}{h}\left(\nabla + \frac{\nabla^2}{2}\right) y_i$$

$$= \frac{1}{2h}(3y_i - 4y_{i-1} + y_{i-2})\,; \; e = O(h^2) \tag{2.36}$$

This result can also be obtained by setting $\nabla = 1 - E^{-1}$ from (2.17) and raising to appropriate powers, that is,

Table 2-2
Relationship Between Backward Difference Operator ∇ and Differential Operator D

	y_i	y_{i-1}	y_{i-2}	y_{i-3}	y_{i-4}	y_{i-5}	*Error*
hDy_i	1	−1					
$h^2D^2y_i$	1	−2	1				$e = O(h)$
$h^3D^3y_i$	1	−3	3	−1			
$h^4D^4y_i$	1	−4	6	−4	1		
$2hDy_i$	3	−4	1				
$h^2D^2y_i$	2	−5	4	−1			
$2h^3D^3y_i$	5	−18	24	−14	3		$e = O(h^2)$
$h^4D^4y_i$	3	−14	26	−24	11	−2	

$$\begin{aligned} Dy_i &= \frac{1}{h}\left(\nabla + \frac{\nabla^2}{2}\right)y_i \\ &= \frac{1}{h}\left(1 - E^{-1} + \frac{(1 - E^{-1})^2}{2}\right)y_i \\ &= \frac{1}{2h}(3y_i - 4y_{i-1} + y_{i-2}) \end{aligned} \tag{2.37}$$

These relations can also be obtained for higher-order differentials. Table 2-2 summarizes the results.

Relationship Between Δ and D

The equations relating forward difference operator Δ and differential operator D are derived in a similar fashion. In this case, from equations (2.17) and (2.22), we have

$$E = 1 + \Delta = e^{hD}$$

$$D = \frac{1}{h}\ln(1 + \Delta) \tag{2.38}$$

resulting in

$$1 + \Delta = 1 + hD + \frac{h^2}{2!}D^2 + \frac{h^3}{3!}D^3 + \ldots$$

$$\Delta = hD + \frac{h^2}{2!}D^2 + \frac{h^3}{3!}D^3 + \ldots \tag{2.39}$$

and

$$D = \frac{1}{h}\left(\Delta - \frac{\Delta^2}{2} + \frac{\Delta^3}{3} - \frac{\Delta^4}{4} + \ldots\right) \tag{2.40}$$

Higher-order derivative and difference operators are obtained by taking various powers of (2.25) and (2.27). This results in the following second- and higher-order operator equations:

$$\Delta^2 = h^2D^2 + h^3D^3 + \frac{7}{12}h^4D^4 + \ldots \tag{2.41}$$

$$\Delta^3 = h^3D^3 + \frac{3}{2}h^4D^4 + \frac{5}{4}h^5D^5 + \ldots \tag{2.42}$$

$$D^2 = \frac{1}{h^2}\left(\Delta^2 - \Delta^3 + \frac{11}{12}\Delta^4 - \frac{5}{6}\Delta^5 + \ldots\right) \tag{2.43}$$

$$D^3 = \frac{1}{h^3}\left(\Delta^3 - \frac{3}{2}\Delta^4 + \frac{7}{4}\Delta^5 - \ldots\right) \tag{2.44}$$

$$D^4 = \frac{1}{h^4}\left(\Delta^4 - 2\Delta^5 + \frac{17}{6}\Delta^6 - \ldots\right) \tag{2.45}$$

Equations (2.40) to (2.45) can be used to find the differentials of functions in terms of available functional values. For example, neglecting errors of order h and higher, D^2y_i can be obtained from (2.43). Dropping Δ^3 and higher-order terms from this equation and using (2.11), we have

$$\begin{aligned} D^2y_i &= \frac{1}{h^2}\Delta^2y_i \\ &= \frac{1}{h^2}(y_{i+2} - 2y_{i+1} + y_i) \end{aligned} \tag{2.46}$$

Using this procedure, a table of coefficients corresponding to Table 2-2 can be calculated for forward-difference operators. This is given in Table 2-3. Note that the second relation above can be derived from the first by substituting $\Delta = 1 - E$ from (2.17) and using the definition of difference operator, that is,

$$\begin{aligned} D^2y_i &= \frac{1}{h^2}(1 - E)^2y_i \\ &= \frac{1}{h^2}(1 - 2E + E^2)y_i \\ &= \frac{1}{h^2}(y_i - 2y_{i+1} + y_{i+2}) \end{aligned} \tag{2.47}$$

Table 2-3
Relationship Between Forward Difference Operator Δ and Differential Operator *D*

	y_i	y_{i+1}	y_{i+2}	y_{i+3}	y_{i+4}	y_{i+5}	*Error*
hDy_i	−1	1					
$h^2D^2y_i$	1	−2	1				
$h^3D^3y_i$	−1	3	−3	1			$e = O(h)$
$h^4D^4y_i$	1	−4	6	−4	1		
$2hDy_i$	−3	4	−1				
$h^2D^2y_i$	2	−5	4	−1			
$2h^3D^2y_i$	−5	18	−24	14	−3		$e = O(h^2)$
$h^4D^4y_i$	3	−14	26	−24	11	−2	

Relationship Between δ and *D* Operators

The equations relating the *central difference operator* δ and differential operator *D* can be obtained using equations (2.17) and (2.22).

$$\delta = E^{1/2} - E^{-1/2},\ E = e^{hD},\ D = \frac{1}{h}\ln E \tag{2.48}$$

In addition, let us define the *mean operator* μ as follows:

$$\mu = \frac{1}{2}(E^{1/2} + E^{-1/2}) \tag{2.49}$$

This will result in

$$\mu\delta = \frac{1}{2}(E - E^{-1}) = \frac{1}{2}\,(e^{hD} - e^{-hD}) = \sinh(hD) \tag{2.50}$$

where the relationship for hyperbolic sine is used in the last expression. Applying the difference operator $\mu\delta$ to y_i, we get,

$$\mu\delta(y_i) = \frac{1}{2}(y_{i+1} - y_{i-1}) \tag{2.51}$$

From (2.49), we have

$$\mu^2 = \frac{1}{4}(E + E^{-1} + 2) = 1 + \frac{\delta^2}{4} \tag{2.52}$$

Substituting the Taylor series expansion for sinh(*hD*), equation (2.50) becomes

$$\mu\delta = hD + \frac{h^3D^3}{6} + \frac{h^5D^5}{120} + \ldots \tag{2.53}$$

Using the expression for δ from (2.48), squaring, and substituting $E = e^{hD}$, we have

$$\delta^2 = (E - 2 + E^{-1}) = 2\left(\frac{e^{hD} + e^{-hD}}{2} - 1\right) \tag{2.54}$$

Substituting the series for $\cosh(hD) = (e^{hD} + e^{-hD})/2$ in (2.54), we get

$$\delta^2 = h^2D^2 + \frac{h^4D^4}{12} + \frac{h^6D^6}{360} + \ldots \tag{2.55}$$

The general expression for δ^n, where $n = 1, 2, 3, \ldots$, can be obtained from equation (2.48).

$$\delta = e^{hD/2} - e^{-hD/2} = 2\sinh\frac{hD}{2}$$

$$\delta^n = 2^n \sinh^n \frac{hD}{2} \tag{2.56}$$

Using combinations of the above relations, we can obtain expressions for higher powers of central difference operators. For example, combination of (2.53) and (2.55) gives

$$\mu\delta^3 = (\mu\delta)(\delta^2) = \left(hD + \frac{h^3D^3}{6} + \frac{h^5D^5}{120} + \ldots\right)$$

$$\left(h^2D^2 + \frac{h^4D^4}{12} + \frac{h^6D^6}{360} + \ldots\right)$$

$$= h^3D^3 + \frac{h^5D^5}{4} + \frac{h^7D^7}{40} + \ldots \tag{2.57}$$

Squaring the first few terms of (2.55) will result in

$$\delta^4 = h^4D^4 + \frac{h^6D^6}{6} + \frac{h^8D^8}{80} + \ldots \tag{2.58}$$

The differential operators D can also be obtained in terms of central difference operators δ.

For example, from (2.50) and (2.53), we get

$$D = \frac{1}{h} \sin h^{-1} \mu\delta = \frac{1}{h}\left(\mu\delta - \frac{\mu\delta^3}{6} + \frac{\mu\delta^5}{30} - \ldots\right) \tag{2.59}$$

where the right side of the equation is obtained from Taylor series expansion of $\sinh^{-1}\mu\delta$. Similarly,

$$D^2 = \frac{1}{h^2}\left(\delta^2 - \frac{\delta^4}{12} + \frac{\delta^6}{90} - \ldots\right) \tag{2.60}$$

$$D^3 = \frac{\mu}{h^3}\left(\delta^3 - \frac{\delta^5}{4} + \frac{7\delta^7}{120} - \ldots\right) \tag{2.61}$$

$$D^4 = \frac{1}{h^4}\left(\delta^4 - \frac{\delta^6}{6} + \frac{7\delta^8}{240} - \ldots\right) \tag{2.62}$$

Note that in these relations only odd differentials (D and D^3) contain terms involving μs.

By the use of the relations, a table can be obtained relating differential and functional values. For example, the second differential D^2, neglecting $e = O(h^2)$, can be obtained from (2.60). From (2.58), δ^4 involves terms of h^4 and higher; since this term (δ^4) in (2.60) is divided by h^2, neglecting it will result in errors of order h^2 and higher. The use of equation (2.13) results in

$$D^2y_i = \frac{1}{h^2}(\delta^2)y_i = \frac{1}{h^2}(y_{i+1} - 2y_i + y_{i-1}); \quad e = O(h^2) \tag{2.63}$$

This result can also be obtained by substituting the value of $\delta = E^{1/2} - E^{-1/2}$ from equation (2.17) and using the definition of operator E, that is

$$\begin{aligned} D^2y_i &= \frac{1}{h^2}(\delta^2)y_i \\ &= \frac{1}{h^2}(E^{1/2} - E^{-1/2})^2y_i \\ &= \frac{1}{h^2}(E - 2 + E^{-1})y_i \\ &= \frac{1}{h^2}(y_{i+1} - 2y_i + y_{i-1}) \end{aligned} \tag{2.64}$$

Table 2-4 gives the relation between various orders of differential operators and central difference operators.

The forward and backward differences are used primarily in incorporating the *initial and/or boundary conditions* in the finite difference representation of the differential equation, while the central differences are used for getting the general finite difference formulation.

Table 2-4
Relationship Between Central Difference Operator δ and Differential Operator *D*

	y_{i-3}	y_{i-2}	y_{i-1}	y_i	y_{i+1}	y_{i+2}	y_{i+3}	Error
$2hDy_i$			-1	0	1			
$h^2D^2y_i$			1	-2	1			
$2h^3D^3y_i$		-1	2	0	-2	1		$e = O(h^2)$
$h^4D^4y_i$		1	-4	6	-4	1		
$12hDy_i$		1	-8	0	8	-1		
$12h^2D^2y_i$		-1	16	-30	16	-1		
$8h^3D^3y_i$	1	-8	13	0	-13	8	-1	$e = O(h^4)$
$6h^4D^4y_i$	-1	12	-39	56	-39	12	-1	

Example: As an example of application of the above equations, consider the numerical solution of the differential equation

$$\frac{d^2y}{dt^2} + \frac{dy}{dt} + y = 0 \tag{2.65}$$

subject to initial conditions at time $t = 0$ of $y = 1$ and $dy/dt = 0$. Using *differential operators* $D = d/dt$ in (2.65), we get

$$D^2y_i + Dy_i + y_i = 0 \tag{2.66}$$

where $D = d/dt$ and the initial conditions ($i = 0$) are $y_0 = 1$, $Dy_0 = 0$. Using the central differences of order h^2, from Table 2-4, we have

$$D^2y_i = \frac{y_{i-1} - 2y_i + y_{i+1}}{h^2}; \quad e = O(h^2)$$

$$Dy_i = \frac{y_{i+1} - y_{i-1}}{2h}; \quad e = O(h^2) \tag{2.67}$$

where $h = \Delta t$ is the assumed time interval. Substituting equation (2.67) in (2.66), we obtain a recursive relation for y_{i+1}, as follows:

$$y_{i+1} = \frac{1}{2 + h}\,[(4 - 2h^2)y_i + (h - 2)y_{i-1}] \tag{2.68}$$

This equation gives the value of y_{i+1} provided that the two previous values of y are available, that is y_i and y_{i-1}. To start the problem, two values of y will be needed. These will be obtained from initial conditions. Using central differences for the initial condition $Dy_0 = 0$, we have

$$Dy_0 = \frac{y_1 - y_{-1}}{2h} = 0; \quad y_{-1} = y_1 \tag{2.69}$$

Note that y_{-1} in the above relation is inserted to make the derivative at y_0 symmetrical. It, y_{-1}, actually does not exist in calculations. In addition, we have the initial conditions give in the problem $y_0 = 1$. Using this condition together with (2.69) and setting $i = 0$ in (2.68), we get

$$y_1 = \frac{1}{2 + h}[(4 - 2h^2)y_0 + (h - 2)y_{-1}]$$

$$= \frac{1}{2 + h}[(4 - 2h^2)(1) + (h - 2)y_1] \quad (2.70)$$

From this

$$y_1 = 1 - \frac{h^2}{2} \quad (2.71)$$

By selecting a value for the increment h, we can obtain a numerical value for y_1. Thus, we will have y_0 and y_1 and we can proceed with the calculation of y_2 by setting $i = 1$ in (2.68). Similarly, we can calculate y_3, y_4 and so on. The values of $y_0, y_1, y_2, y_3, \ldots$ will represent the solution of the differential equation (2.65) at discrete time intervals of $i\Delta t = ih$ where $i = 0, 1, 2, 3, \ldots$

Mathematically, one of the important problems in the numerical solution of differential equations is choosing the value $h = \Delta t$. If Δt is too small (in addition to the problems of the stability of solution which may arise), a great number of iterations will be required to obtain the value of y for a given time. For example, if the value of y at $t = 10$ is required, a $\Delta t = 0.01$ will require 1,000 applications of equation (2.68), while a $\Delta t = 0.1$ will require only 100 applications of this equation to obtain the desired value of y. On the other hand, if Δt is too large, since the error is of the order of Δt^2 [or h^2 in equation (2.67)], the error in each application of (2.68) may accumulate, resulting in large errors in the solution.

In most engineering applications, however, the value of Δt is chosen by numerical experiments. This means that a value of Δt is chosen, the problem run on the computer, and solutions obtained. The value of Δt is then varied and the solutions examined. In this way, a value is chosen that is a trade-off between computation time and the desired accuracy of the solution.

Integration Formulas by Interpolating Polynomials

The formulas already given and additional integration formulas for the numerical solution of differential equations can be obtained by using polynomial interpolations. The basic idea is that a polynomial with unknown coefficients is assumed to pass through the sets of given points. Using the values of these points, the coefficients of the polynomial can be evaluated. These coefficients are then used to obtain integration formulas.

Consider Figure 2-1 with variables x and $y = f(x)$. Assume that the straight line

$$y = Ax + B \tag{2.72}$$

passes through (x_i, y_i) and (x_{i+1}, y_{i+1}). For simplicity, let us set $x_i = 0$ or equivalently take x_i as the starting point and set the points $(0, y_i)$ and $(\Delta x, y_{i+1})$ in equation (2.72). This results in

$$\begin{aligned} y_i &= B \\ y_{i+1} &= A(\Delta x) + B \end{aligned} \tag{2.73}$$

Solving these equations for A and B, we obtain

$$\begin{aligned} A &= \frac{y_{i+1} - y_i}{\Delta x} \\ B &= y_i \end{aligned} \tag{2.74}$$

Note that the value of A is identical to the forward difference representation of order h given for the first derivative operator in Table 2-3, the reason being that the first derivative of (2.72) with respect to x is also equal to A. Thus, the forward difference equation represents the derivative of the function obtained by passing a line through the two points of the function.

Using the values of A and B from (2.74) in (2.72) we get,

$$y = \frac{y_{i+1} - y_i}{\Delta x} x + y_i \tag{2.75}$$

The area under the curve from $x = 0$ to $x = \Delta x$ is obtained by integration of equation (2.75). This gives

$$\begin{aligned} A &= \int_0^{\Delta x} y dx = \frac{y_{i+1} - y_i}{\Delta x} \left(\frac{\Delta x^2}{2}\right) + y_i \Delta x \\ &= \frac{\Delta x}{2}(y_{i+1} + y_i) \end{aligned} \tag{2.76}$$

which is the *trapezoidal* rule of integration. In a similar fashion, the Simpson rule of integration can be obtained by assuming the second-degree polynomial

$$y = Ax^2 + Bx + C \tag{2.77}$$

passing through the points $(-\Delta x, y_{i-1})$, $(0, y_i)$, $(\Delta x, y_{i+1})$. Substituting these values in (2.77) and solving for A, B, and C, we obtain

$$A = \frac{y_{i+1} - 2y_i + y_{i-1}}{2(\Delta x)^2}, \quad B = \frac{y_{i+1} - y_{i-1}}{2\Delta x}, \quad C = y_i \tag{2.78}$$

Using these values of A, B, and C in (2.77) and integrating.

$$\text{Area} = \int_{-\Delta x}^{\Delta x} y dx = \frac{\Delta x}{3}(y_{i+1} + 4y_i + y_{i-1}) \tag{2.79}$$

Equation (2.79) represents the *Simpson's one-third* rule of integration.

Expressions (2.76) and (2.79) give areas under the curve between points which satisfy expressions (2.72) and (2.77). It is sometimes desirable, as will be seen in *predictor-corrector* methods of solving differential equations, to extend the range of integration using the same polynomials. For example, using (2.77) with the coefficients (2.78), the area under the curve of Figure 2-1 from $-2\Delta x$ to $2\Delta x$ becomes

$$\int_{-2\Delta x}^{2\Delta x} y dx = \frac{4\Delta x}{3}(2y_{i+1} - y_i + 2y_{i-1}) \tag{2.80}$$

This expression gives the area under the curve extending from (x_{i-2}, y_{i-2}) to (x_{i+2}, y_{i+2}) in terms of values of y extending from y_{i-1} to y_{i+1}.

Differentiating (2.77) twice will result in

$$\frac{d^2y}{dx^2} = 2A \tag{2.81}$$

Substituting the value of A from equation (2.78), we obtain

$$\frac{d^2y}{dx^2} = \frac{y_{i+1} - 2y_i + y_{i-1}}{\Delta x^2} \tag{2.82}$$

which is the *central difference* expression for the second derivative of y at x_i, as seen in Table 2-4.

Differentiating (2.77) once, we obtain

$$\frac{dy}{dx} = 2Ax + B \tag{2.83}$$

The value of this derivative at $x = x_i = 0$ is obtained by substituting B from (2.78) and $x = 0$. This results in

$$\frac{dy}{dx} = \frac{y_{i+1} - y_{i-1}}{2\Delta x} \tag{2.84}$$

which is the *central difference* expression for the first derivative.

From the above formulations it is seen that the central differences of order h^2 for the first and the second derivatives given in Table 2-4 can also be obtained by passing a second-degree polynomial through three discrete points.

The integrals derived here are used in the numerical solution of differential equations.

Example: Formulate the solution of the boundary-value differential equation

$$\frac{d^2y}{dx^2} + 3\frac{dy}{dx} + 6y = x^2 + 3 \tag{2.85}$$

with the boundary conditions

$$\begin{cases} x = 0 \\ y = 2 \end{cases} \quad \text{and} \quad \begin{cases} x = 10 \\ dy/dx = 3 \end{cases} \tag{2.86}$$

Note the difference between *initial* and *boundary* conditions. *Initial conditions* refer to cases where the values of the dependent variable (y) and its derivatives are given at the initial point ($x = 0$) of the independent variable. *Boundary conditions* refer to cases where the values of the dependent variable and/or its derivatives are given at the boundaries—points where the independent variable is not necessarily zero. Replacing the differential equation with the central difference relations of order h^2 from Table 2-4, we have

$$\frac{y_{i+1} - 2y_i + y_{i-1}}{\Delta x^2} + 3\frac{y_{i+1} - y_{i-1}}{2\Delta x} + 6y_i = (i\Delta x)^2 + 3 \tag{2.87}$$

where the value of x on the right side of the equation is replaced by $i\Delta x$.

Assuming the interval $x = 0$ to $x = 10$ is divided into N steps of Δx each, we have

$$\Delta x = \frac{10}{N} \tag{2.88}$$

From the boundary condition at $x = 0$, we have $y = 2$. The boundary condition at $x = 10$ may be represented by the backward difference relation from Table 2-2.

$$Dy_N = \frac{dy_N}{dx} = \frac{y_N - 4y_{N-1} + y_{N-2}}{2\Delta x} = 3 \tag{2.89}$$

where the last expression involves an error of order h^2.

Setting the value of index $i = 1$ in equation (2.87), we get the first equation of the set,

$$y_1[6\Delta x^2 - 2] + y_2\left[1 + \frac{3}{2}\Delta x\right] = (\Delta x)^4 + 3\Delta x^2 + 3\Delta x - 2 \quad (2.90)$$

where y_0 is set at 2.0. Using $i = 2, 3, \ldots, N-1$ in equation (2.87) will give the rest of the equations with the exception of the last one, which is given by equation (2.89). Thus, the set of N simultaneous equations in N unknowns $y_1, y_2, \ldots, y_N$ can be solved for the values of independent variable y. Note that the values $y_1, y_2, \ldots, y_N$ represent the variable y at discrete intervals Δx, $2\Delta x$, $3\Delta x$, $\ldots$, $N\Delta x$.

Problems

1. Derive the relation
   $\nabla^3 y_i = y_i - 3y_{i-1} + 3y_{i-2} - y_{i-3}$
2. Derive the relation
   $\Delta^3 y_i = y_{i+3} - 3y_{i+2} + 3y_{i+1} - y_i$
3. Complete the values of the Problem Table 2A

Problem Table 2A
Difference Operators of $y = x^3$

i	x_i	y_i	∇y_i	$\nabla^2 y_i$	$\nabla^3 y_i$	$\nabla^4 y_i$
1	0					
2	1					
3	2					
4	3					
5	4					
6	5					
7	6					
8	7					
9	8					
10	9					

4. Complete the values of the Problem Table 2B

Problem Table 2B
Difference Operators of $y = 2x^3 + 3$

i	x_i	y_i	Δy_i	$\Delta^2 y_i$	$\Delta^3 y_i$	$\Delta^4 y_i$
1	0					
2	1					
3	2					
4	3					
5	4					
6	5					
7	6					
8	7					
9	8					
10	9					

5. Complete the values of the Problem Table 2C

Problem Table 2C
Difference Operators of $y = 2x^2 + 1$

i	x_i	y_i	$\mu\delta y_i$	$\delta^2 y_i$	$\mu\delta^3 y_i$ $\mu\delta(\delta^2 y_i)$
1	0				
2	1				
3	2				
4	3				
5	4				
6	5				
7	6				
8	7				
9	8				
10	9				

6. Formulate the solution of the differential equation

$$\frac{d^2y}{dx^2} + 2\frac{dy}{dx} + 3y = 8x$$

with the initial conditions

$$x = 0, \quad y = 1$$

$$x = 0, \quad \frac{dy}{dx} = 6$$

using central differences.

7. Formulate the solution of the differential equation

$$\frac{d^2y}{dx^2} + 3x^2y = 6x$$

with the initial conditions

$$x = 0, \quad y = 1$$

$$x = 0, \quad \frac{dy}{dx} = 0$$

using central differences.

8. Formulate the solution of boundary value problem

$$\frac{d^2y}{dx^2} + x\frac{dy}{dx} + y = x^3$$

with the boundary conditions

$$x = 0, \quad y = 1$$

$$x = 5, \quad y = 10$$

using central differences.

9. Formulate the solution of boundary value problem

$$\frac{d^2y}{dx^2} + x^2\frac{dy}{dx} + y = 6$$

with the boundary conditions

$$x = 0, \quad \frac{dy}{dx} = 6$$

$$x = 5, \quad \frac{dy}{dx} = 10$$

using central, forward, and backward differences.

10. Using polynomial fit of the type:

$$y = Ax^2 + Bx + C$$

derive the integration formula

$$\int_{-2\Delta x}^{2\Delta x} y\,dx = \frac{4\Delta x}{3}(2y_{i+1} - y_i + 2y_{i-1})$$

11. Take the equation

$$y = Ax^2 + Bx + C$$

and assume it passes through the points $(0, y_i)$, $(\Delta x, y_{i+1})$ and $(2\,\Delta x, y_{i+2})$. Evaluate the coefficients A, B, and C and obtain

$$\text{Area} = \int_0^{2\Delta} y\,dx$$

12. Same as problem 11, but this time calculate an extended area

$$\text{Area} = \int_{-\Delta x}^{3\Delta x} y\,dx$$

13. The Gregory-Newton interpolation formula is derived by the following considerations. Consider the function $f(x)$ tabulated from $x = a$ to $x = b$ at increments of $\Delta x = h$. It is desired to obtain $f(x)$ at an intermediate value of $x = a + nh$ where n can be a fraction. Using the relation

$$\begin{aligned} f(a + nh) &= E^n f(a) \\ &= (1 + \Delta)^n f(a) \end{aligned}$$

obtain the Gregory-Newton formula in terms of forward differences.:

$$f(a + nh) = \left[1 + n\Delta + \frac{n(n-1)}{2!}\Delta^2 + \frac{n(n-1)(n-2)}{3!}\Delta^3 + \ldots\right] f(a)$$

14. Using the procedure of problem 13 and the relation

$$\begin{aligned} E^{-n} f(b) &= f(b - nh) \\ &= (1 - \nabla)^n f(b) \end{aligned}$$

obtain the Gregory-Newton interpolation formula in terms of backward differences:

$$f(b - nh) = \left[1 - n\nabla + \frac{n(n-1)}{2!}\nabla^2 - \frac{n(n-1)(n-2)}{3!}\nabla^3 + \ldots\right] f(b)$$

15. Consider the table of $\log_{10} x$ for values of x between 1.0 and 1.05 to be

given at increments $\Delta x = h = 0.01$. Find the value of $\log x$ at $x = 1.024$ by the forward interpolation formula of problem 13.

x	$\log_{10} x$
1.00	0.0000000
1.01	0.0043214
1.02	0.0086002
1.03	0.0128372
1.04	0.0170333
1.05	0.0211893

16. The Sterling interpolation formula is based on central differences. In the derivation of this formula, consider the Taylor series:

$$f(a + nh) = \left(1 + nhD + \frac{n^2h^2}{2!} D^2 + \frac{n^3h^3}{3!} D^3 + \dots\right) f(a)$$

Replacing the derivative terms with central differences, obtain the Sterling interpolation formula:

$$f(a + nh) = \left[1 + n\mu\delta + \frac{n^2}{2!}\delta^2 + \frac{n(n^2 - 1)}{3!} \mu\delta^3 + \frac{n^2(n^2 - 1)}{4!}\delta^4 + \dots\right] f(a)$$

17. Using the first four terms of the Sterling interpolation formula of problem 16, obtain the value of log 1.024. The tabulated values of log x are given in problem 15.
18. Using the tabulated values of tan x for $x = 0.50$ to 0.60 in intervals of $\Delta x = 0.02$

x	tan x
0.50	0.54630
0.52	0.57256
0.54	0.59943
0.56	0.62695
0.58	0.65517
0.60	0.68414

find the values of tan 0.55 using the Sterling formula of problem 16.
19. Obtain the first and second derivative of tan x at $n = 0.54$, using the values given in problem 18 and the central difference formulas of Table 2-4 for error of order h^4. Compare these values to values obtained by differentiating tan x.
20. Using the tabulated values of tan x of problem 18, obtain the area under the curve between $n = 0.5$ and $n = 0.58$ using trapezoidal and Simpson's one-third rule (central difference formulation).

Note

1. W.E. Boyce and R.C. DiPrima, *Elementary Differential Equations*, (New York: Wiley, 1969).

Bibliography

Hamming, R.W. *Numerical Methods for Scientists and Engineers*. New York: McGraw-Hill, 1973.

Hildebrand, F.B. *Methods of Applied Mathematics*. Englewood Cliffs, N.J.: Prentice-Hall, 1963.

Hovanessian, S.A., and Pipes, Louis A., *Digital Computer Methods in Engineering*. New York: McGraw-Hill, 1969.

Pipes, L.A. *Applied Mathematics for Engineers and Physicists*. New York: McGraw-Hill, 1958, chapter 10.

3 Numerical Solution of Ordinary Differential Equation

The classical solution of ordinary differential equations results in the values of the dependent variable y as a function of the independent variable x; i.e., $y = f(x)$. This solution can be evaluated at any x by setting the value of x in $f(x)$ and obtaining a numerical value for $y = f(x)$.

On the other hand, the numerical solution of ordinary differential equations involves the selection of an increment Δx and the evaluation of the solution at each increment Δx, $2\Delta x$, $3\Delta x$, and so on. Thus, the solution at $i\Delta x$, where i is an integer, will depend on the previously evaluated solutions $(i - 1)\Delta x$, $(i - 2)\Delta x$, and so on. Having obtained the values of the dependent variable y at increments of Δx, a plot of y vs. x can be made, representing the solution of the differential equation.

This chapter will discuss some preliminary notions of numerical solutions, starting from the Taylor series expansion. We will then extend this to several commonly used numerical formulations for the solution of differential equations. The discussion will include Runge-Kutta methods and predictor-corrector methods.

Preliminary Discussion

Consider the first-order differential equation

$$y' = dy/dx = f(x, y) \tag{3.1}$$

with the initial condition $y = y_0$ at $x = x_0$. We write the Taylor series expansion of y about x_0, using an increment Δx.

$$y(x_0 + \Delta x) = y_0 + \Delta x y_0' + \frac{1}{2!}(\Delta x)^2 y_0'' + \ldots \tag{3.2}$$

Taking the first two terms of the above series, we get (see Figure 3-1)

$$\begin{aligned} y(x_0 + \Delta x) &= y_0 + \Delta x y'_0 \\ &= y_0 + \Delta x f(x_0, y_0) \end{aligned} \tag{3.3}$$

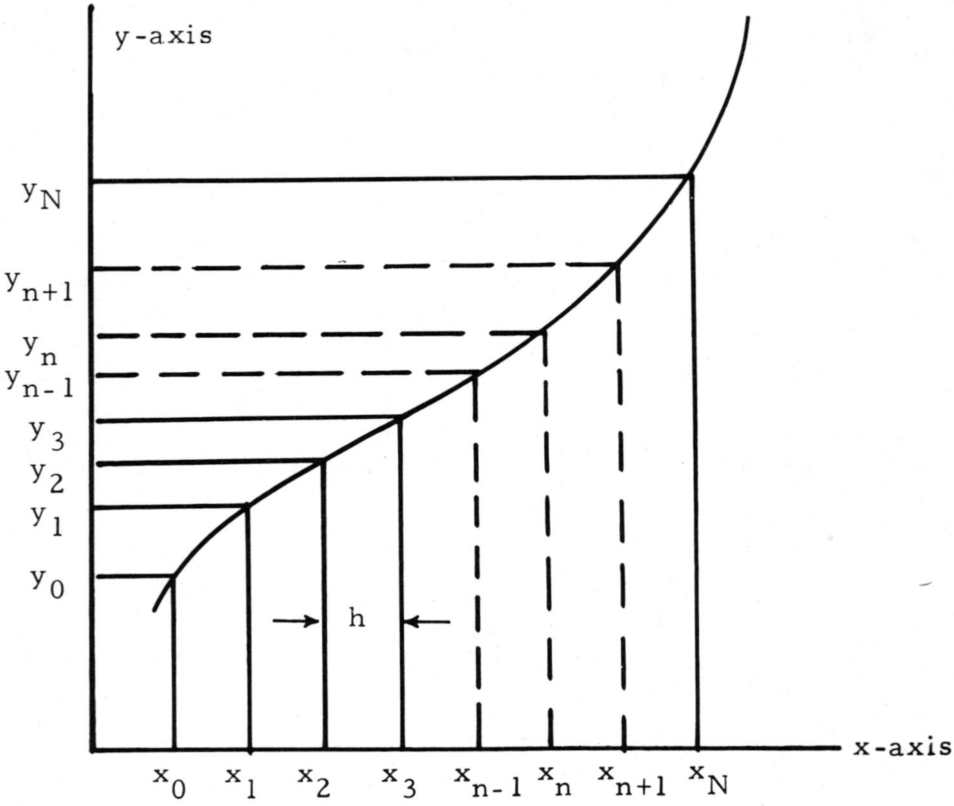

Figure 3-1. Function y vs. x with Increments of h

Denoting $x_0 + \Delta x = x_1$ and $x_2 = x_1 + \Delta x$ with the corresponding $y(x_0 + \Delta x) = y_1$ and $y(x_1 + \Delta x) = y_2$, we may write

$$y_1 = y_0 + \Delta x f(x_0, y_0)$$

$$y_2 = y_1 + \Delta x f(x_1, y_1) \tag{3.4}$$

$$\cdots$$

$$y_{n+1} = y_n + \Delta x f(x_n, y_n) \tag{3.5}$$

Since we used only the first two terms of the Taylor series expansion (3.2), we may proceed to obtain the magnitude of the third term $1/2!\,(\Delta x)^2 y''$. This term will represent the order of magnitude of the error introduced in representation (3.5).

We may compute the total derivative of equation (3.1) as follows,

$$dy' = \frac{\partial f}{\partial x}dx + \frac{\partial f}{\partial y}dy \tag{3.6}$$

where $\partial f/\partial x$ and $\partial f/\partial y$ represent partial derivatives of $f(x, y)$ with respect to x and y. Dividing the above by dx, we get

$$\begin{aligned} \frac{dy'}{dx} &= \frac{\partial f}{\partial x} + \frac{\partial f}{\partial y}\frac{dy}{dx} \\ &= \frac{\partial f}{\partial x} + f\frac{\partial f}{\partial y} \qquad (3.7) \\ &= y'' \end{aligned}$$

Writing the Taylor series expansion (3.2) for $y_{n+1} = y(x_n + \Delta x)$ and using the above expression, we get

$$y_{n+1} = y_n + \Delta x f(x_n, y_n) + \frac{1}{2}(\Delta x)^2\left[\frac{\partial f}{\partial x} + f\frac{\partial f}{\partial y}\right]_n + \ldots \tag{3.8}$$

Comparing this with equation (3.5), we obtain the error term e.

$$e = \frac{\Delta x^2}{2}\left(\frac{\partial f}{\partial x} + f\frac{\partial f}{\partial y}\right); \tag{3.9}$$

Note that the introduced error is directly proportional to the square of the increment Δx and, additionally, it is proportional to the combination of function $f(x, y)$ and its partial derivatives with respect to x and y. These partial derivatives are a measure of the rate of change of $f(x, y)$ with respect to x and y.

Example: Consider the numerical solution of the following first-order differential equation:

$$\frac{dy}{dx} = y' = -y \quad \text{with} \quad y = 1 \quad \text{at} \quad x = 0 \tag{3.10}$$

To obtain the analytical solution, we note that the derivative of the function y satisfying the above differential equation is equal to the negative of the function itself. An exponential function with a negative exponent will satisfy this condition, so that

$$y = Ae^{-x} \tag{3.11}$$

where A is a constant to be evaluated using the condition $y = 1$ at $x = 0$. Thus, $A = 1$ and the solution becomes

$$y = e^{-x} \tag{3.12}$$

We may use the finite difference representation given in Chapter 2 to write the difference equations representing the differential equation or we may proceed with the following intuitive reasoning.

The derivative of y at x_n is equal to the slope of the curve of Figure 3-1 at this point. This slope can be represented by either one of the formulations

$$y'_n = (y_{n+1} - y_n)/h$$

$$y'_n = (y_{n+1} - y_{n-1})/2h \tag{3.13}$$

where h is equal to the increment Δx. They can be recognized as the forward and central difference formulations of the first derivative. Noting that the differential equation requires the slope y' to be equal to the negative of the function y, we may write

$$(y_{n+1} - y_n)/h = -y_n; \quad y_0 = 1, \; h = \frac{1}{4} \tag{3.14}$$

$$(y_{n+1} - y_{n-1})/2h = -y_n; \quad y_0 = 1, \; y_1 = e^{-1/4}, \; h = \frac{1}{4} \tag{3.15}$$

$$(y_{n+1} - y_{n-1})/2h = -y_n; \quad y_0 = 1, \; y_1 = e^{-1/2}, \; h = \frac{1}{2} \tag{3.16}$$

Formulation (3.14) is the forward difference formulation with increment $h = 0.25$. Formulations (3.15) and (3.16) are both central difference formulations with $h = 0.25$ and $h = = 0.50$, respectively. We can set $n = 0$ in formulation (3.14) and solve for y_1. Having obtained y_1, we set $n = 1$ and obtain y_2 and so on.

Formulation (3.15), however, will require both y_0 and y_1 to start the problem and solve for y_2. Setting $n = 1$ in this formulation will give y_2 in terms of y_0 and y_1. The value of y_1 needed here (in addition to y_0) is calculated from the analytical solution for illustration purposes. Note that in practical problems the analytical solution is not usually available and y_1 cannot be evaluated in this manner. It can, however, be evaluated by using the Taylor expansion series. Solution (3.16) is also a central difference formulation similar to (3.15), but with a larger increment; $h = 0.50$ instead of 0.25.

Analytical solution (3.12) together with finite difference solutions (3.14), (3.15) and (3.16) are evaluated and are given in Table 3-1. This ta-

Table 3-1
Solutions of the Example Problem

	Equation			
x	*(3.12)*	*(3.14)*	*(3.15)*	*(3.16)*
0.0	1.00000	1.00000	1.00000	1.00000
		0.75000	0.77880	
0.5	0.60653	0.56250	0.61060	0.60653
		0.42188	0.47350	
1.0	0.36788	0.31641	0.37385	0.39347
		0.23730	0.28658	
1.5	0.22313	0.17798	0.23056	0.21306
		0.13348	0.17130	
2.0	0.13534	0.10011	0.14491	0.18041
		0.07508	0.09885	
2.5	0.08209	0.05631	0.09549	0.03265
		0.04224	0.05111	
3.0	0.04979	0.03168	0.06994	0.14776
		0.02376	0.01614	
3.5	0.03020	0.01782	0.06187	−0.11511
		0.01336	−0.01480	
4.0	0.01832	0.01002	+0.06927	+0.26287
		0.00752	−0.04944	
4.5	0.01111	0.00564	+0.09399	−0.37798
		0.00423	−0.09644	
5.0	0.00674	0.00317	+0.14221	+0.64085
		0.00238	−0.16754	
5.5	0.00409	0.00178	+0.22598	−1.01883

ble shows that only numerical solution (3.14), which is a forward difference formulation with $h = 0.25$, seems to be stable and follows the analytical solution (3.12). The accuracy of this solution can be improved by reducing h, e.g., to 0.125. Solutions (3.15) and (3.16) seem to be unstable with fluctuating signs and do not follow along the lines of the analytical solution (3.12).

This example was given to point out the fact that formulas with greater accuracy, such as the central difference formulation with error of order h^2, do not necessarily produce the correct results. Note that the accuracy of the forward difference formulation producing the correct result was of order h.

Intuitively, the problem encountered with solutions (3.15) and (3.16) might have been anticipated had we considered the following reasoning. The central difference representation of the first derivative is

$$y' = \frac{y_{n+1} - y_{n-1}}{2h} \tag{3.17}$$

and that of the second derivative is

$$y'' = \frac{y_{n+1} - 2y_n + y_{n-1}}{h^2} \tag{3.18}$$

Both formulas are of order h^2. It is also noted that the range of values of y in both formulas is the same, y_{n-1} to y_{n+1}. Thus, by using a central difference formula for y' in the differential equation, without the presence of y'' term, we are forcing, in effect, a second-order difference representation. This second-order difference representation, as will be seen in the discussion of stability of solution given at the end of this chapter, will result in an oscillatory solution. In general, however, when we have second-order differentials present in the differential equation, the central difference formulation of order h^2 results in satisfactory solutions.

Analytical methods for the study of stability of numerical formulations will be given at the end of this chapter. As will be seen at that time, application of this method in practical problems is extremely difficult.

Series Expansion

In order to start a number of numerical formulations, such as formulations (3.15) and (3.16) of the previous problem and the predictor-corrector methods to be discussed, one usually needs several initial values of the dependent variable. For example, one may need y_0, y_1, y_2, and y_3 evaluated at 0, Δx, $2\Delta x$, and $3\Delta x$ in order to use some of the numerical formulations. To obtain these values with reasonable accuracy, one may apply the Taylor series expansion for their evaluation. This series, however, can be evaluated only if the derivatives of the dependent variable y can be calculated.

As discussed perviously, the Taylor series of $y(x + \Delta x)$ can be written as a function of $y(x)$ and its derivatives, as

$$y(x + \Delta x) = y(x) + \Delta x y'(x) + \frac{\Delta x^2}{2!} y''(x) + \frac{\Delta x^3}{3!} y'''(x) + \dots \tag{3.19}$$

Given the value of $y(x)$ and its derivatives, the value of $y(x + \Delta x)$ can be computed by using the preceding series. The procedure then can be repeated to obtain $y(x + 2\Delta x)$, $y(x + 3\Delta x)$, and so on.

Example: Consider the example of the first-order differential equation, where prime denotes the derivative with respect to x:

$$y' = 1 - 2xy \tag{3.20}$$

with the condition $y = 0$ at $x = 1$. It is required to compute y at $x = 1$ using four terms of the Taylor series expansion. First, we compute derivatives y', y'', and y''' at $x = 1$ and $y = 0$,

$$\begin{aligned} y' &= 1 - 2xy = 1.0 \\ y'' &= -2xy' - 2y = -2.0 \\ y''' &= -2xy'' - 4y' = 0.0 \end{aligned} \tag{3.21}$$

Setting these values in the Taylor series expansion, we get

$$\begin{aligned} y &= 0 + (0.1)(1.0) + \frac{1}{2!}(0.1)^2(-2.0) + \frac{1}{3!}(0.2)^3(0) \\ &= 0 + 0.1 + 0.01 + 0.0 = 0.11 \end{aligned} \tag{3.22}$$

Having the value of y at $x = 1.1$, we can rewrite the Taylor series expansion around $x = 1.1$ and compute the value of y at $x = 1.2$, and so on. The computed values of y at $x = 1.1, 1.2, 1.3, \ldots$ can be used as starting values needed in various formulations or they can constitute the solution of the problem by themselves.

Runge-Kutta Methods

The Runge-Kutta methods are among the most highly used and recommended methods in the solution of ordinary differential equations. In addition to their accuracy, they do not need several values of the dependent variable for starting the solution.

There are several formulations of Runge-Kutta methods, each derived by making assumptions for constants encountered in their derivation. The following is a brief description of the procedure involved in deriving Runge-Kutta formulas.

Consider the solution of the differential equation $y' = f(x, y)$ with the initial conditions (x_0, y_0) and an increment of $\Delta x = h$. We set

$$\begin{aligned} k_1 &= hf(x_0. y_0) \\ k_2 &= hf(x_0 + mh, y_0 + mk_1) \\ k_3 &= hf(x_0 + nh, y_0 + rk_2 + (n - r)k_1) \\ k_4 &= hf(x_0 + ph, y_0 + sk_2 + tk_3 + (p - s - t)k_1) \\ k &= ak_1 + bk_2 + ck_3 + dk_4 \end{aligned} \tag{3.23}$$

The constants $m, n, r, p, s, \ldots$ should be determined in such a way as to make $y_0 + k$ approach $y(x_0 + h)$.

By use of series expansions and some complicated mathematics, one obtains

$$\begin{aligned} a + b + c + d &= 1 & cmr + d(nt + ms) &= 1/6 \\ bm + cn + dp &= 1/2 & cmnr + dp(nt + ms) &= 1/8 \\ bm^2 + cn^2 + dp^2 &= 1/3 & cm^2 + d(n^2t + m^2s) &= 1/2 \\ bm^3 + cn^3 + dp^3 &= 1/4 & dmrt &= 1/24 \end{aligned} \tag{3.24}$$

We have eight equations and ten unknowns. Selecting $m = n$ and $b = c$, we get

$$\begin{aligned} m = n &= 1/2 & s &= 0 & a = d &= 1/6 \\ p &= 1 & t &= 1 & b = c &= 1/3 \\ r &= 1/2 \end{aligned} \tag{3.25}$$

Thus we have the formulas

$$\begin{aligned} k_1 &= hf(x_0, y_0) \\ k_2 &= hf(x_0 + 1/2h, y_0 + 1/2k_1) \\ k_3 &= hf(x_0 + 1/2h, y_0 + 1/2k_2) \\ k_4 &= hf(x_0 + h, y_0 + k_3) \end{aligned} \tag{3.26}$$

and

$$k = 1/6\,(k_1 + 2k_2 + 2k_3 + k_4)$$

The formula has an error of order h^4. Note that having x_0, y_0 and selecting h, k_1 of equation (3.26), can be computed. Using this, k_2 can be computed and, subsequently, k_3 and k_4. The increment of y corresponding to increment h of x is computed from the last of equations (3.26). Thus,

$$y(x + h) = y + k \tag{3.27}$$

Having obtained $y(x + h)$, one can reapply (3.26) and obtain $y(x + 2h)$, and so on.

Several sets of Runge-Kutta formulas are available in the literature. These formulas are all derived by using (3.24) and making varying assumptions in the selection of two of the ten unknowns discussed earlier.[1]

Extension to Sets of Equations

Consider the two first-order differential equations, where primes denote derivatives with respect to x.

$$y' = F(x, y, z)$$
$$z' = G(x, y, z) \tag{3.28}$$

Runge-Kutta formulas can be written for this set as

$$k_1 = hF(x, y, z)$$
$$k_2 = hF(x + h/2, y + k_1/2, z + l_1/2)$$
$$k_3 = hF(x + h/2, y + k_2/2, z + l_2/2)$$
$$k_4 = hF(x + h, y + k_3, z + l_3)$$
$$k = 1/6\,(k_1 + 2k_2 + 2k_3 + k_4)$$
$$\tag{3.29}$$
$$l_1 = hG(x, y, z)$$
$$l_2 = hG(x + h/2, y + k_1/2, z + l_1/2)$$
$$l_3 = hG(x + h/2, y + k_2/2, z + l_2/2)$$
$$l_4 = hG(x + h, y + k_3, z + l_3)$$
$$l = 1/6(l_1 + 2l_2 + 2l_3 + l_4)$$

These formulas will result in new values $(x + h, y + k, z + l)$ where h is the selected increment. Equations (3.29) are similar to previously given equations (3.26). Selection of a $\Delta x = h$ increment allows one to proceed with the calculations of k_1, l_1, k_2, l_2, . . ., and, finally, k and l. The k and l increments are used to obtain new y and z values.

Numerical Example: Consider the solution of the second-order differential equation

$$y'' = xy'^2 - y^2 \tag{3.30}$$

with initial conditions $x = 0$, $y = 1$, $y' = 0$. Primes denote derivatives with respect to x. To apply the Runge-Kutta method, the second-order differential equation is reduced to two first-order equations by defining the new variable z, where $z = dy/dx = y'$. Using this in the differential equation, we have

$$y' = z = F(x, y, z)$$
$$z' = xz^2 - y^2 = G(x, y, z) \tag{3.31}$$

where F and G are newly defined functions.

Assume that we want to compute y and z for $x = 0.20$, given the initial conditions at $x = 0$. Using equations (3.29) with $h = 0.20$, we get

$$\begin{aligned} k_1 &= 0.20\,F(0,\ 1,\ 0) = 0 & l_1 &= 0.20\,G(0,\ 1,\ 0) = -0.20 \\ k_2 &= -0.02 & l_2 &= -0.1998 \\ k_3 &= -0.01998 & l_3 &= -0.1958204 \\ k_4 &= -0.039164 & l_4 &= -0.1905542 \end{aligned} \tag{3.32}$$

Using the above values to compute k and l from the last of equations (3.29), we get

$$k = -0.019854 \quad l = -0.196966 \tag{3.33}$$

Then for $x = 0.2$, we get

$$\begin{aligned} y &= 1 + k = 0.980146 \\ z &= 0 + l = -0.196966 \end{aligned} \tag{3.34}$$

These values may constitute the values of the dependent variables at the next step. With these, the same procedure can be carried out to compute y and z for $x = 0.4, 0.6, 0.8$, and so on.

Milne's Predictor-Corrector Methods

Predictor-corrector methods are based upon two integration formulas. One formula is used to predict the solution for the next time increment Δt or space increment Δx, and the other formula is used to correct the calculated value. The method is described for a first-order differential equation:

$$y' = f(x, y) \tag{3.35}$$

where prime denotes the derivative with respect to x.

The predictor formulas generally use integrals with extended integration range, such as the one derived in Chapter 2 that is given:

$$\int_{-2h}^{2h} y\,dx = \frac{4h}{3}(2y_{-1} - y_0 + 2y_1) + 0(h^5)y^V \tag{3.36}$$

The range of integration is from $-2h$ to $2h$ or over four $\Delta x = h$ increments. Note also that the right side of the equation contains values of y at $x = -\Delta x, 0, \Delta x$ and *not* at $2\Delta x$ and $-2\Delta x$. Thus, the integral extends beyond the range of y values. The error term of the equation is proportional to h^5 multiplied by the fifth derivative of y with respect to x.

Substituting the function y'_n for y, we get

$$y_{n+2} - y_{n-2} = \frac{4h}{3}(2y'_{n-1} - y'_n + 2y'_{n+1}) + 0(h^5)y^V \qquad (3.37)$$

Shifting the index of n by one unit, i.e., replacing n by $n - 1$, and solving for y_{n+1}, we get the predictor formula

$$y_{n+1} = y_{n-3} + \frac{4h}{3}(2y'_{n-2} - y'_{n-1} + 2y'_n) \text{ (predictor)} \qquad (3.38)$$

The predictor formula gives the value of y_{n+1} in terms of known quantities y_{n-3}, y'_{n-2}, y'_{n-1}, y'_n up to and including the nth step. Having the value of y_{n+1} from the above, we can calculate

$$y'_{n+1} = f(x_{n+1}, y_{n+1}) \qquad (3.39)$$

using the given differential equation.

Now we can use the corrector formula, which gives the value of y_{n+1} in terms of currently available values of y and y' up to and including the $(n + 1)$th step.[2]

$$y_{n+1} = y_{n-1} + \frac{h}{3}(y'_{n-1} - 4y'_n + y'_{n+1})$$

$$+ \bar{0}(h^5)y^V(\zeta) \text{ (corrector)} \qquad (3.40)$$

Having the above corrected value of y_{n+1}, we can apply the corrector formula again by computing y'_{n+1} from (3.39). Usually, however, one application of the corrector formula is sufficient, as will be seen in the following example. Having obtained y_{n+1}, we can again start with the predictor formula and compute y_{n+2}, and so on.

The predictor-corrector methods can be mechanized (hard wired) by using digital circuit components. Studies of this kind come under the general heading of digital filters. Because of current interest in this type of mechanization, the equivalent digital circuit of the predictor-corrector method described is given in Appendix 3A.

Numerical Example: Consider the differential equation

$$x' = f(x, t) = t^3 + x \qquad (3.41)$$

where prime denotes derivative with respect to time t. Initial conditions are $x = 1$ at $t = 1$. Let us also assume for the purposes of comparison that the analytical solution of the differential equation is given as

$$x = 17e^{t-1} - t^3 - 3t^2 - 6t - 6 \qquad (3.42)$$

Table 3-2
Results of Predictor-Corrector Method

Time[a]	*Predicted* x^a	*Corrected* x^a	*Function*[a]	*Analytical Solution*
1.0	1	1	2	1
1.1	1.22691	1.22691	2.55791	1.22691
1.2	1.51585	1.51585	3.24385	1.51585
1.3	1.88060	1.88060	4.07760	1.88060
1.4	2.33696	2.33702	5.08102	2.33702
1.5	2.90320	2.90326	6.27826	2.90326
1.6	3.59994	3.60002	7.69602	3.60002
1.7	4.45071	4.45080	9.36380	4.45080
1.8	5.48210	5.48220	11.3142	5.48220
1.9	6.72415	6.72426	13.5833	6.72425
2.0	8.21068	8.21080	16.2108	8.21079
2.1	9.97970	9.97983	19.2408	9.97982
2.2	12.0739	12.0740	22.7220	12.0740
2.3	14.5409	14.5411	26.7081	14.5410
2.4	17.4342	17.4344	31.2584	17.4344
2.5	20.8135	20.8137	36.4387	20.8137
2.6	24.7454	24.7456	42.3216	24.7456
2.7	29.3039	29.3041	48.9871	29.3041
2.8	34.5718	34.5720	56.5240	34.5720
2.9	40.6409	40.6412	65.0302	40.6412
3.0	47.6137	47.6140	74.6140	47.6140
3.1	55.6036	55.6039	85.3949	55.6039

[a]The first four values are computed from Taylor series expansion.

For a $\Delta t = 0.10$, the predictor-corrector method will also require values of x at $t = 1.0,\ 1.1,\ 1.2$, and 1.3. Assume these initial values are computed by using Taylor series expansion as given in Table 3-2. Having the values of x at Δt increments, we can compute the *function* $f(x, t) = t^3 + x$ as given in the table. Having x_0 thru x_3 and f_0 through f_3, we can write the *predictor* formula (3.38)

$$x_4 = x_0 + \frac{4\Delta t}{3}(2f_3 - f_2 + 2f_1) = 2.33696 \tag{3.43}$$

Having x_4 and $t_4 = 1.4$, we can compute $f_4 = 5.08096$. Now the corrector formula is applied:

$$x_4 = x_2 + \frac{\Delta t}{3}(f_2 + 4f_3 + f_4) = 2.33702 \tag{3.44}$$

The process can be continued with the tabulated results given in Table 3-2.

From Table 3-2, we see a very small difference between predicted and corrected values. We should note, however, that the calculation of the predicted values was based upon the use of the corrected values of x in the previous step. Thus, the predicted values are only prediction for one time-step duration and thereby are not much different from the corrected values. Another possible comparison would have been the case where the corrector formula (3.40) was not used at all, and all of the calculations were made using the predictor formula (3.38).

Numerov's Method

For second-order differential equations of the type

$$y'' = F(x, y) \qquad (3.45)$$

where the first derivative with respect to x, (y') does not appear in the function $F(x, y)$, Numerov's method gives an accurate solution. Differential equations of this type occur in a number of engineering applications, such as vibration analysis and the wave equation.

Consider equidistance points along x with separation distance of h and assume y_n, y_{n-1} and F_n, F_{n-1} are known. The general form of the finite difference representation of (3.45) can be written

$$y_{n+1} + Ah^2y''_{n+1} = By_n + Ch^2y''_n + Dy_{n-1} + Eh^2y''_{n-1} + R \qquad (3.46)$$

where A, B, C, D, and E are constants and R represents the error term. Numerov's method for the evaluation of the constants consists of writing the Taylor series expansion of y_{n+1} and equating the terms of this expansion to those of equation (3.46). This analysis, after considerable manipulation, results in the values of the constants:

$$A = -1/12, \quad B = 2, \quad C = 5/6,$$

$$D = -1, \text{ and } E = 1/12 \qquad (3.47)$$

The remainder term becomes

$$R = -(h^6/240)y_n^{VI} \qquad (3.48)$$

The remainder is proportional to the product of the sixth power of increment h and the sixth derivative y_n^{VI}. Note that the usual representation of a second-order differential equation by central differences produces an error of order h^2.

Using the above constants in (3.46), we get

$$y_{n+1} - \frac{h^2}{12}F_{n+1} = 2y_n + \frac{5h^2}{6}F_n - y_{n-1} + \frac{h^2}{12}F_{n-1}$$

$$+ \text{ERROR } 0\left(\frac{h^6}{240}\right)y_n^{VI} \qquad (3.49)$$

The above equation, without the error term, can best be solved by iteration, as follows:

1. Neglect F_{n+1} and solve for y_{n+1}.
2. Calculate F_{n+1} using new y_{n+1}.
3. Calculate a new y_{n+1} using F_{n+1}.

In the particular case where $F(x, y)$ of differential equation (3.45) can be factored in the form

$$F(x, y) = y \cdot f(x) \tag{3.50}$$

where $f(x)$ is a function of x only, equation (3.49) becomes

$$y_{n+1} = \frac{[2 + (5/6)h^2 f_n]\, y_n - [1 - (h^2/12)\, f_{n-1}]\, y_{n-1}}{1 - (h^2/12)\, f_{n+1}} \tag{3.51}$$

Computations can proceed as outlined above, using equation (3.51) instead of (3.49).

Boundary Value Problems

Consider the differential equation

$$y'' + \frac{y}{1 + x^2} = 7x \tag{3.52}$$

with the boundary condition

$$\begin{cases} y = 0 \\ x = 0 \end{cases} \qquad \begin{cases} y = 2 \\ x = 1 \end{cases} \tag{3.53}$$

Note that the above relations specify the values of the function y at the end points $x = 0$ and $x = 1$. It is required to find the values of y at the intermediate values of $0 < x < 1$. The initial value problem corresponding to (3.52) will specify conditions of the function and its derivative (y and y') at $x = 0$.

The numerical solution of (3.52) will be different from that of the corresponding initial value problem, as the final solution should include the condition at the boundary $x = 1$. There are two commonly used methods for the solution of boundary value problems: simultaneous equations and trial-and-error methods. In the simultaneous equation method, the boundary value differential equation is replaced by a series of finite-difference representations at each increment of the independent variable. Referring to equation (3.52), the finite-difference representation of the last increment incorporates the second boundary condition at $x = 1$. Thus, in this method, the solution of the boundary value problem is reduced to the solution of a set of linear simultaneous equations with the unknowns being the values of the dependent variable at the chosen increments.

The second method of solving boundary value problems, the trial-and-error method, replaces the differential equation by a finite difference equivalent, similar to the simultaneous equation method. Using one of the boundary conditions (at $x = 0$) and estimating the value of the dependent variable at the next increment, y_1 at Δx, this finite-difference equation is solved progressively for the dependent variables $y_2, y_3, \ldots y_n$. This solution will result in a value for the independent variable y_n at the second boundary point ($x = 1.0$). If this value matches with the given boundary condition, the solution is terminated. If not, a new estimate for the value of y_1 is made and the procedure repeated, resulting in a new value of y_n at the boundary $x = 1.0$. At this point, we will have two calculated values for the second boundary, which can be used for estimating a new value of y_1, continuing the solution by trial and error. Application of the trial-and-error procedure to the solution of boundary value problems, in effect, reduces the problem to an initial value one. The following examples will illustrate the procedures.

Simultaneous Equation Method

Replace the differential equation (3.52) by the difference equation

$$\frac{y_{n+1} - 2y_n + y_{n-1}}{h^2} + \frac{y_n}{1 + x^2} = 7x \tag{3.54}$$

where h is the increment in x. The first term is the central difference representation of y''. Using $h = 0.2$, we get

$$y_{n+1} - \frac{1.96 + 2x^2}{1 + x^2} y_n + y_{n-1} = 0.28x \tag{3.55}$$

Using the notation $y(0.2) = y_1$, $y(0.4) = y_2$, $y(0.6) = y_3$, and $y(0.8) = y_4$, we can write the set of equations

$$\begin{aligned}
-1.9615y_1 + y_2 &= 0.056 \\
y_1 - 1.9655y_2 + y_3 &= 0.112 \\
y_2 - 1.9706y_3 + y_4 &= 0.168 \\
y_3 - 1.9756y_4 &= -1.776
\end{aligned} \tag{3.56}$$

Note that the boundary condition at $x = 1.0$ or $y(1.0) = y_5 = 2.0$ is incorporated in the last of equations (3.56). The solution of the set of equations (3.56) is obtained using the matrix methods described in Chapter 1. This solution is:

$$y_1 = 0.208$$

$$y_2 = 0.464$$

$$y_3 = 0.816$$

$$y_4 = 1.312 \tag{3.57}$$

This solution agrees with the analytical solution $y = x^3 + x$. It is noted from the example that the numerical solution of the boundary value differential equation results in a set of simultaneous equations to be solved by previously given methods.

Trial-and-Error Method

The solution of the boundary value differential equation (3.52) can also be obtained by the trial-and-error method previously outlined. Again, let us select h as the increment of the independent variable x and write (3.52) in the finite-difference form similar to (3.54):

$$\frac{y_{n+1} - 2y_n + y_{n-1}}{h^2} + \frac{y_n}{1 + (x_n)^2} = 7x_n \tag{3.58}$$

Replacing x_n in (3.58) by its equivalent value (i.e., $x_n = nh$) and solving for y_{n+1}, we get

$$y_{n+1} = y_n\left[2 - \frac{h^2}{1 + (nh)^2}\right] - y_{n-1} + h^2(7nh) \tag{3.59}$$

Using the numerical values of $n = 1$ and $h = 0.2$, (3.59) becomes

$$y_2 = 1.9615y_1 - y_0 + 0.056 \tag{3.60}$$

From the boundary condition at $x = 0$, we have $y_0 = 0$. We need an estimate, however, for the value of y_1 to start the trial-and-error procedure. Since we have five space increments for $h = 0.2$, extrapolating the boundary conditions (3.53) linearly, we get $y_1 = 0.4$. Using this value in (3.60), we get

$$y_2 = 0.8406 \tag{3.61}$$

Having the values of $y_1 = 0.4$ and $y_2 = 0.8406$, we can set $n = 2$ in (3.59) and solve for y_3. This will result in a $y_3 = 1.3642$. Similarly, we can get the values of y_4 and y_5 from (3.59) as

$$y_4 = 2.0157\,; \quad y_5 = 2.8420 \tag{3.62}$$

We note that the value of y_5 in (3.62) is larger than the corresponding value of y_5 from boundary conditions as given in (3.53). Now we may reduce the estimate of y_1 from its previous value of 0.4 and proceed with the calculations. Selecting a new value of $y = 0.2$, equation (3.59) for $n = 1$, $h = 0.2$, and $y_0 = 0$ results in

$$y_2 = 0.4483 \tag{3.63}$$

Using the values of $y_1 = 0.2$ and $y_2 = 0.4483$ with $n = 2$ and $h = 0.2$, equation (3.59) results in $y_3 = 0.7931$. Similarly, we get $y_4 = 1.2825$ and $y_5 = 1.9646$. This value of y_5 is very close to the corresponding boundary value of 2.0 given in (3.53). Linear extrapolation of estimates of y_1 and the corresponding y_5 values will give a new estimate for the value of y_1, as follows:

$$\begin{aligned} y_1 &= 0.4 - \frac{(0.4 - 0.2)}{(2.8420 - 1.9646)}(2.8420 - 2.0) \\ &= 0.2081 \end{aligned} \tag{3.64}$$

Using this value and following the outlined procedure starting from equation (3.59), we get

$$\begin{aligned} y_2 &= 0.4642 \\ y_3 &= 0.8163 \\ y_4 &= 1.3124 \\ y_5 &= 2.0005 \end{aligned} \tag{3.65}$$

Note that the value of y_5 in (3.65) is very close to the boundary condition value of 2.0 given in (3.53), and the trial-and-error procedure can be terminated.

The trial-and-error method works best for linear differential equations. The presence of nonlinearities may slow down the process of convergence of the solution, extending the number of trials required. It is seen that the number of arithmetical operations in using the trial-and-error method depends on the convergence rate of the solution. The number of operations in using the simultaneous equation method depends on the method used for solving these equations and can usually be estimated in advance. It should also be pointed out that simultaneous equation methods usually require a large computer memory for the storage of the coefficient matrix while iterative methods do not.

Solution by Direct Substitution

Differential equations arising in servo systems and in a number of other engineering applications are usually solved by the method of direct substitution. In this method, the differential equation is replaced by a finite-difference equation and solution proceeds along the lines discussed earlier. This section will illustrate the application of the method to two initial-value differential equations. The second example is given to illustrate also the reduction of higher order differential equations to sets of first-order differential equations.

Consider the set of differential equations to be solved by the method of direct substitution.

$$\frac{dy}{dt} = 3y + t^2 - z$$

$$\frac{dz}{dt} = y + 1 - t^2 + z \tag{3.66}$$

with $t = 0$, $y = 1$, $z = 1$. Find the solution for $t = 0$ using a $\Delta t = 0.10$. Consider the following definitions:

$$\begin{cases} x_1 = y \\ x_2 = z \end{cases} \tag{3.67}$$

resulting in the sets of equations

$$\begin{cases} \dot{x}_1 = 3x_1 + t^2 - x_2 \\ \dot{x}_2 = x_1 + 1 - t^2 + x_2 \end{cases} \tag{3.68}$$

or, in matrix notation,

$$\dot{\mathbf{X}} = \mathbf{AX} + \mathbf{B} \text{ where } \mathbf{X} = \begin{pmatrix} x_1 \\ x_2 \end{pmatrix}; \mathbf{A} = \begin{pmatrix} 3 & -1 \\ 1 & 1 \end{pmatrix} \tag{3.69}$$

and

$$\mathbf{B} = \begin{pmatrix} t^2 \\ 1 - t^2 \end{pmatrix} \tag{3.70}$$

Using the forward difference representation for **X**,

$$\dot{\mathbf{X}} = \frac{\mathbf{X}_{n+1} - \mathbf{X}_n}{\Delta t} \tag{3.71}$$

we get

$$\mathbf{X}_1 = \begin{bmatrix} 1.3 & -0.1 \\ 0.1 & 1.1 \end{bmatrix} \begin{pmatrix} 1 \\ 1 \end{pmatrix} + \begin{pmatrix} 0 \\ 0.1 \end{pmatrix} = \begin{pmatrix} 1.2 \\ 1.3 \end{pmatrix}$$

$$\mathbf{X}_2 = \begin{pmatrix} 1.3 & -0.1 \\ 0.1 & 1.1 \end{pmatrix} \begin{pmatrix} 1.2 \\ 1.3 \end{pmatrix} + \begin{pmatrix} 0.001 \\ 0.099 \end{pmatrix} = \begin{pmatrix} 1.431 \\ 1.649 \end{pmatrix}$$

$$\cdots$$

$$\mathbf{X}_{10} = \begin{pmatrix} 1.3 & -0.1 \\ 0.1 & 1.1 \end{pmatrix} \begin{pmatrix} 4.6599 \\ 6.1614 \end{pmatrix} + \begin{pmatrix} 0.081 \\ 0.019 \end{pmatrix} = \begin{pmatrix} 5.5227 \\ 7.2625 \end{pmatrix} \tag{3.72}$$

Thus, $y(1) = 5.5227$ and $z(1) = 7.2625$. The iteration process (3.72) can easily be programmed for digital computer solution.

Another Example

Consider the set of second-order differential equation

$$\frac{d^2x}{dt^2} + 2\left(\frac{dy}{dt} - 3\right) = 0$$

$$\frac{d^2y}{dt^2} - 2\left(\frac{dx}{dt} - 5\right) = 0 \tag{3.73}$$

$$\text{at } t = 0 \quad \begin{cases} x = 1 \\ \dot{x} = 0 \\ y = 1 \\ \dot{y} = 0 \end{cases}$$

Define $x_1 = x$, $x_2 = y$, $x_3 = \dot{x}$, $x_4 = \dot{y}$ and obtain

$$\dot{\mathbf{X}} = \begin{pmatrix} \dot{x}_1 \\ \dot{x}_2 \\ \dot{x}_3 \\ \dot{x}_4 \end{pmatrix} = \begin{bmatrix} 0 & 0 & 1 & 0 \\ 0 & 0 & 0 & 1 \\ 0 & 0 & 0 & -2 \\ 0 & 0 & 2 & 0 \end{bmatrix} \begin{pmatrix} x_1 \\ x_2 \\ x_3 \\ x_4 \end{pmatrix} + \begin{pmatrix} 0 \\ 0 \\ 6 \\ -10 \end{pmatrix} \tag{3.74}$$

or, using matrix notation,

$$\dot{\mathbf{X}} = \mathbf{AX} + \mathbf{B} \tag{3.75}$$

Replacing $\dot{\mathbf{X}}_n$ by the forward difference equivalent $(\mathbf{X}_{n+1} - \mathbf{X}_n)/\Delta t$ and solving for $\mathbf{X}_{n+1}$, we get

$$\mathbf{X}_{n+1} = (\mathbf{A}\Delta t + \mathbf{I})\mathbf{X}_n + \mathbf{B}\Delta t \tag{3.76}$$

where **I** is the unit matrix. Equation (3.76) in expanded form becomes

$$\mathbf{X}_{n+1} = \left\{ \begin{bmatrix} 0 & 0 & 1 & 0 \\ 0 & 0 & 0 & 1 \\ 0 & 0 & 0 & -2 \\ 0 & 0 & 2 & 0 \end{bmatrix} \Delta t + \begin{bmatrix} 1 & 0 & 0 & 0 \\ 0 & 1 & 0 & 0 \\ 0 & 0 & 1 & 0 \\ 0 & 0 & 0 & 1 \end{bmatrix} \right\} \begin{pmatrix} x_1 \\ x_2 \\ x_3 \\ x_4 \end{pmatrix}$$

$$+ \begin{pmatrix} 0 \\ 0 \\ 6 \\ -10 \end{pmatrix} \Delta t \tag{3.77}$$

Using $\Delta t = 0.1$, we get, where primes denote transpose matrices,

$$\mathbf{X}_0 = (1,\ 1,\ 0,\ 0)'$$

$$\mathbf{X}_1 = (1,\ 1,\ 0.6,\ -1.0)'$$

$$\cdots$$

$$\mathbf{X}_5 = (1.7872,\ 0.1395,\ 4.7209,\ 3.4256)' \text{ at } t = 0.5 \tag{3.78}$$

and at $t = 1$, we have $x_1 = x = 5.4182$, $x_2 = y = -1.320$.

As in the case of the previous example, equation (3.77) can be programmed for digital computer solution.

Simulation of Servo Systems

Consider the servo system of Figure 3-2, where s denotes Laplace transform notation which is equivalent to the derivative notation d/dt. We are given the input to the servo as a function of time, as shown in Figure 3-3. We want to find the output V of the servo as a function of time, possibly as shown in Figure 3-4.

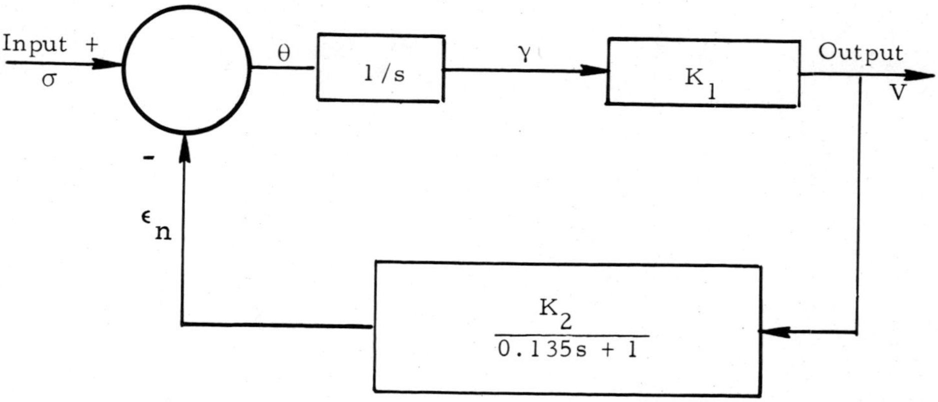

Figure 3-2. Servo System of the Example Problem

Solution: The servo loop of Figure 3-2 can be replaced by the following sets of equations, written in logical order

$$\theta_n = \sigma_n - \epsilon_n$$

$$\frac{\theta}{s} = \gamma \quad \text{or} \quad \theta = s\gamma$$

$$\theta_n = \frac{d\gamma}{dt} = \frac{\gamma_{n+1} - \gamma_n}{T} \tag{3.79}$$

$$\gamma_{n+1} = T\theta_n + \gamma_n$$

$$V_{n+1} = K_1\gamma_{n+1}$$

where T is the time-step size. The following additional equations can also be written:

$$\epsilon = \frac{K_2V}{0.135s + 1}$$

$$0.135s\epsilon + \epsilon = K_2V$$

$$0.135\,\frac{\epsilon_{n+1} - \epsilon_n}{T} + \epsilon_n = K_2V_{n+1}$$

$$\epsilon_{n+1} = \epsilon_n + \frac{T}{0.135}(K_2V_{n+1} - \epsilon_n) \tag{3.80}$$

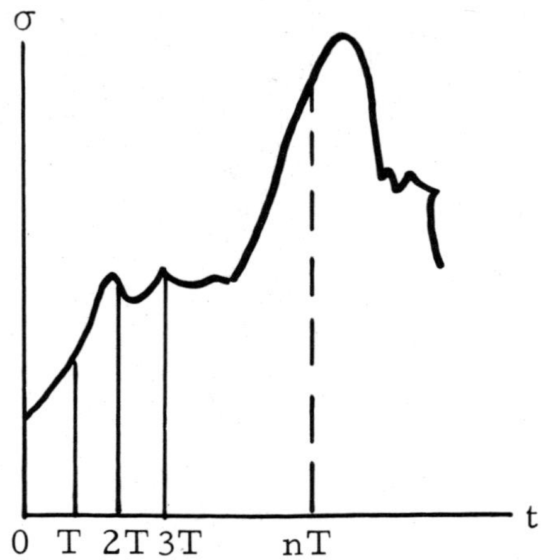

Figure 3-3. Input to Servo System

Figure 3-4. Possible Output of the Servo System

We are given values of the input function at time steps 0, 1, 2, . . . as σ_0, σ_1, σ_2, ... Using initial values of $\epsilon_0 = 0$ and $\gamma_0 = 0$, we intend to find the values of the output V at time steps 1, 2, 3,. . . . These are written as V_1, V_2, V_3, ... For various consecutive values of time step n, we have

For $n = 0$

$$\theta_0 = \sigma_0$$

$$\gamma_1 = T\theta_0$$

$$V_1 = K_1\gamma_1$$

$$\epsilon_1 = \epsilon_0 + \frac{T}{0.135}(K_2V_1) \tag{3.81}$$

For $n = 1$

$$\theta_1 = \sigma_1 - \epsilon_1$$

$$\gamma_2 = T\theta_1 - \gamma_1$$

$$V_2 = K_1\gamma_2$$

$$\epsilon_2 = \epsilon_1 + \frac{T}{0.135}(K_2V_2 - \epsilon_1) \tag{3.82}$$

$$\cdots$$

For $n = i$

$$\theta_i = \sigma_i - \epsilon_i$$

$$\gamma_{i+1} = T\theta_i - \gamma_i$$

$$V_{i+1} = K_1\gamma_{i+1}$$

$$\epsilon_{i+1} = \epsilon_i + \frac{T}{0.135}(K_2V_{i+1} - \epsilon_i) \tag{3.83}$$

Finally, the solution can be obtained and plotted as given in Figure 3-4. Thus, the numerical solution, or the digital computer simulation, of servo systems is reduced to writing the equivalent differential equations in numerical form and solving these in order for each time step 0, Δt, 2Δ,. . . .

Appendix 3A gives the mechanization of these equations by the use of digital components.

Stability of the Solution of Difference Formulations

The stability of solution of difference formulations is analyzed through the use of differential-difference equations.[3] The solutions of the differential-difference equations parallel those of differential equations as given in books on this topic. The stability analysis, however, can be performed for cases where analytical solutions are available. At first sight, this might be a shortcoming of this analysis, since analytical solutions are not to be found for many practical problems. On the other hand, practical problems can usually be simplified to cases where solutions are available. The stability analysis of these simplified problems may give insight for the engineer to use in the solution of practical cases. In this section, a brief description of the method is given. Following this, the method is applied to an example problem, solution of which was given in Table 3-1.

The finite difference formulations result in equations of the type

$$x(t + iT) + v_1 x(t + (i - 1)T) + \ldots + v_i x(t) = f(t) \tag{3.84}$$

Using the notation

$$E^i x(t) = x(t + iT) \tag{3.85}$$

the above equation becomes

$$(E^i + v_1 E^{i-1} + \ldots + v_i)x(t) = f(t) \tag{3.86}$$

The total solution of the difference equation $x(t)$ can be represented as the sum of the complementary solution $x_c(t)$ and the particular solution $x_p(t)$. The *complementary* solution is obtained by setting $f(t) = 0$. Denoting the complementary solution by $x_c(t) = Ce^{mt}$ (The corresponding analysis can be reviewed in any differential equations text book), we get

$$Ce^{mt}(e^{miT} + v_1 e^{m(i-1)T} + \ldots + v_i) = 0 \tag{3.87}$$

or denoting $q = e^{mT}$, we get

$$q^i + v_i q^{i-1} + v_2 q^{i-2} + \ldots + v_i = 0 \tag{3.88}$$

Assuming $t = nT$, we get

$$x_c(nT) = Ce^{mnT} = Cq^n \tag{3.89}$$

where qs are the roots of equation (3.88). If the roots are *distinct*

$$x_c(nT) = C_1 q_1^n + C_2 q_2^n + \ldots + C_i q_i^n \tag{3.90}$$

where Cs are used as a general constant throughout. If the roots are complex conjugate,

$$q_1 = a + bj; \quad q_2 = a - bj \tag{3.91}$$

$$x_c(nT) = R^n(C_1 \cos n\theta + C_2 \sin n\theta) \tag{3.92}$$

where

$$R = \sqrt{a^2 + b^2}; \quad \theta = \tan^{-1}(b/a) \tag{3.93}$$

In the case of repeated roots, $q_1 = q_2 = q$

$$x_c(nT) = q^n(C_1 + C_2 n) \tag{3.94}$$

The particular solution $x_p(nT)$ is obtained from an examination of deriving function $f(t)$. The particular solution will be of the form

1. $f(t) = A^{kt} = A^{knT}$; $\quad x_p(nT) = CA^{knT}$ (3.95)

2. $f(t) = At^m = A(nT)^m$; $\quad x_p(nT) = C_1(nT)^m + C_2(nT)^{m-1} + \ldots + C_{m+1}$ (3.96)

3. $f(t) = \sin At$ or $\cos At$ or

   $f(t) = \sin AnT$ or $\cos Ant$ (3.97)

   Then

$$x_p(nT) = C_1 \sin AnT + C_2 \cos Ant$$

These equations are used in examining the stability of solution of a given numerical formulation.

Example: Consider the solution of $y' = -y$ subject to $y = 1$ at $x = 0$, with the following finite difference formulations (exact solution $y = e^{-x}$)

$$\frac{y_{n+1} - y_n}{h} = -y_n; \quad y_0 = 1,\ h = 0.25 \tag{3.98}$$

and

$$\frac{y_{n+1} - y_{n-1}}{2h} = -y_n; \quad y_0 = 1,\ y_1 = e^{-0.25}, \quad h = 0.25 \tag{3.99}$$

For solution (3.98) we have

$$y_{n+1} - y_n(1 - h) = 0 \tag{3.100}$$

or

$$q - (1 - h) = 0; \quad q = 1 - h = 0.75 \tag{3.101}$$

The complementary solution becomes

$$y_n = C(0.75)^n \tag{3.102}$$

Using $y_0 = 1$ when $n = 0$, we get

$$y_n = (0.75)^n \tag{3.103}$$

For solution (3.99), we get

$$y_{n+1} + 2hy_n - y_{n-1} = 0 \tag{3.104}$$

or

$$y_{n+2} + 2hy_{n+1} - y_n = 0$$

$$(E^2 + 2hE - 1)y_n = 0 \tag{3.105}$$

or

$$q^2 + 2hq - 1 = 0$$

$$q = -h \pm \sqrt{h^2 + 1} = 0.7808 \text{ and } -1.2808 \tag{3.106}$$

Then

$$y_n = C_1(0.7808)^n + C_2(-1.2808)^n \tag{3.107}$$

Using $y_0 = 1$ when $n = 0$ and $y_1 = e^{-0.25} = 0.77880$ when $n = 1$, we get

$$y_n = 0.99904(0.7808)^n + 0.00096(-1.2808)^n \tag{3.108}$$

It is seen that solution (3.103) is monotonically decreasing while solution (3.108) is oscillatory. Since the actual solution of the problem is $y = e^{-x}$, the oscillatory solution and the corresponding formulation (3.99) are not acceptable.

Discussion of the Methods

Several methods for the numerical solution of differential equations were outlined in this chapter. The simplest method is the method of direct substitution whereby the differential equation is replaced by a finite difference approximation. This method, however, may give rise to errors if the finite difference formulation is not stable and/or the time-step selection is not appropriate. The former was illustrated by formulation (3.99), while the latter was illustrated by solutons given in Table 3-1.

It is seen that stability analysis can be applied to primarily linear differential equations for which analytical solutions are available. This analysis should be used as a guide in the solution of nonlinear differential equations for which analytical solutions are not available. In these cases, the problem is usually linearized and the finite difference equations are tested for stability. If the finite difference formulation is *not* stable for the linearized equation, its application to the case of the *nonlinear* system may give rise to serious errors.

Usually, a particular formulation for the solution of differential equations is tested for a number of input conditions for which the behavior of the system may be known either from physical data or analysis. These checks develop confidence in the formulation and allow one to extend the solution to the cases with varying input parameters. Time-step selections are also examined during this process. After multiplication or division of the iteration time step by two, the answers are examined against the requirements of the simulation until an appropirate time step is reached. Note that a time step greater than required may use large amounts of computer time unnecessarily.

In engineering applications, it is rare that the formulation of a system or a problem will result in a single differential equation or a set of differential equations. However, if this case arises, one may use any of the methods—Runge-Kutta, Milne's prediction-corrector, Numerov's, and so on—for a more exact solution.

The more common case in engineering applicaions, however, is where a differential equation is to be solved in conjunction with the physical simulation of a system. In these cases, the method of direct substitution outlined in equations (3.79) and (3.80) is probably the most appropriate. In the case of servo loops, for example, conditions change at each iteration time step Δt. This requires the numerical formulation to yield a correct answer, based on the values at the beginning of the time step, for only one iteration step. This, together with the fact that servo systems are error-compensating, allows one to select a simpler numerical formulation for the encountered differential equations.

Problems

1. Consider the differential equation with initial conditions as follows:

$$\frac{d^2y}{dx^2} + 7\frac{dy}{dx} + 2y = 16$$

   with $y = 1$, $dy/dx = 2$ at $x = 0$. Obtain the value of y at $x = 0.1$ and 0.2 by evaluating four terms of the Taylor series expansion.

2. Consider the differential equation

$$\frac{d^2y}{dx^2} + 3\frac{dy}{dx} + 8y = 3x$$

   with $y = 3$, $dy/dx = 6$ at $x = 1$. Obtain y at $x = 1.1$ and 1.2. Evaluate four terms of the Taylor series expansion.

3. In problem 2, evaluate y' at $x = 1.1$ by writing the four terms of the Taylor series expansion of y'.

4. Consider the differential equation

$$\frac{d^2y}{dx^2} + 3\sin x = x^2$$

with $y = 1$, $dy/dx = 6$ at $x = 2$. Evaluate y at $x = 2.1$, 2.2, and 2.3 by using four terms of the Taylor series expansion.

5. Represent the differential equation of problem 1 by central differences and obtain the solution y for $x = 0.1, 0.2$, and 0.3 with $\Delta x = 0.1$.
6. Represent the differential equation of problem 2 by central differences and obtain the solution y at $x = 1.1, 1.2, 1.3, 1.4$, and 1.5 with $\Delta x = 0.1$.
7. Reduce the second-order differential equations of problems 1 and 2 to two first-order differential equations.
8. Using Runge-Kutta formulas, solve the differential equation

$$\frac{d^2y}{dx^2} + 6\frac{dy}{dx} + 3y = x^2 + x$$

with $y = 1$, $dy/dx = 2$ at $x = 0$. Obtain the solution at $x = 0.1, 0.2$, and 0.3 with $\Delta x = 0.1$.

9. In problem 8 obtain the solution at $x = 0.2$, using $\Delta x = 0.2$.
10. Reduce the third-order differential equation

$$\frac{d^3y}{dt^3} + 3\frac{d^2y}{dt^2} + 6\frac{dy}{dt} + 7y = t^2 + 6$$

into three first-order differential equations. Incorporate the initial conditions $y = 1$, $dy/dt = 0$, $d^2y/dt^2 = 6$ at $t = 0$.

11. Using Milne's predictor-corrector method, obtain the values given in Table 3-2 for time equal to 1.4.
12. Formulate the solution of the problem

$$\frac{dy}{dt} + 3y = 6x^2$$

with $y = 1$ at $x = 0$, by Milne's predictor-corrector method.

13. Formulate the solution of problem 1 by Milne's predictor-corrector method.
14. Formulate the solution of problem 2 by Milne's predictor-corrector method.
15. Using Numerov's method, solve the differential equation

$$\frac{d^2x}{dt^2} + x = 0$$

with the initial conditions $t = 0$, $x = 1.0$, and $dx/dt = 0$. Assume the first few points are evaluated by Taylor series expansion as

$$t = 0 \qquad x = 1.0$$
$$t = 0.10 \qquad x = 0.995$$
$$t = 0.20 \qquad x = 0.980$$

Using $\Delta t = 0.1$, solve for values of x at $t = 0.3$, 0.4, and 0.5.

16. Formulate the solution of differential equation

$$\frac{d^2y}{dx^2} = x^2 + 3y$$

with the initial conditions $x = 1$, $y = 1$, $dy/dx = 0$ by Numerov's method.

17. Formulate the solution of the boundary value problem

$$\frac{d^2y}{dx^2} + 3\frac{dy}{dx} + y = x^2$$

with $y = 1$ at $x = 0$ and $y = 6$ at $x = 2$, using central difference representation.

18. Formulate the solution of the boundary value problem

$$2\frac{d^2y}{dx^2} + \frac{dy}{dx} + y = 12$$

with $y = 1$ at $x = 0$ and $y = 2$ at $x = 1$, using central difference formulations. Assume the segment $x = 0$ to 1 is divided into 10 divisions.

19. Formulate the solution of the boundary value problem

$$\frac{d^2y}{dx^2} + \frac{dy}{dx} + 6y = x^3$$

with $y = 1$ at $x = 0$ and $dy/dx = 2$ at $x = 1$. Use central differences to represent the differential equation and backward differences to represent the boundary condition. Assume the segment $x = 0$ to 1 is divided into 10 divisions.

20. Formulate the solution of differential equation

$$\frac{d^2y}{dx^2} + 3\frac{dy}{dx} + y = x$$

with $dy/dx = 2$ at $x = 0$ and $dy/dx = 6$ at $x = 5$. Use central differences to represent the differential equation and use forward and backward differences to represent the boundary conditions. Assume the segment $x = 0$ to 5 is divided into 10 divisions.

21. Formulate the solution of differential equation

$$\frac{d^3y}{dt^3} + 2t^2 \frac{d^2y}{dt^2} + 2ty - 10 = 0$$

with $t = 0$, $y = 1$, $\dot{y} = 1$, and $\ddot{y} = 0$. Reduce this equation to a set of three simultaneous equations.

22. Formulate the solution of the set of equations

$$\frac{dx}{dt} = x^2 + t$$

$$\frac{d^2y}{dt^2} = x + y - t$$

with $t = 0$, $x = 1$, $y = 0$, $\dot{y} = 1$. Obtain numerical values of x and y at $t = 0.1$, 0.2, and 0.3 with $\Delta t = 0.1$.

23. Formulate the possible digital simulation of the servo system

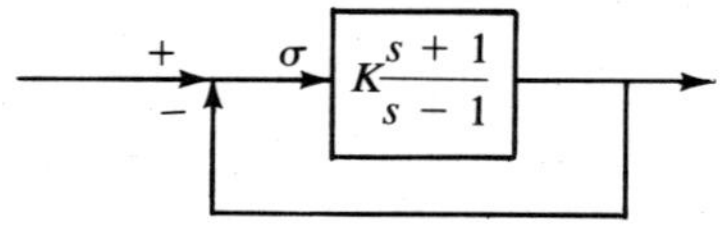

24. Formulate the possible digital computer simulation of

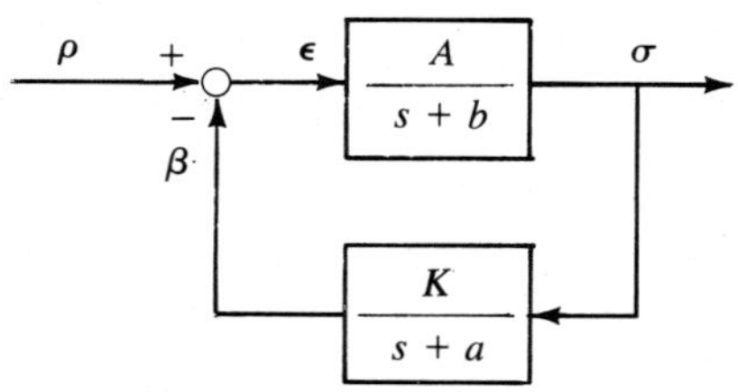

25. Consider the problem

$$\frac{d^2x}{dt^2} + \omega^2 x = 0$$

with the initial conditions $t = 0$, $x = 1$, and $dx/dt = 0$. The analytic solution is given

$$x(t) = \cos n(\omega T)$$

where T is the time step and n is the iteration number. Consider the numerical formulation

$$\frac{x_{n+1} - 2x_n + x_{n-1}}{T^2} + \omega^2 x_n = 0$$

Show that the solution becomes oscillatory only for $\omega T < 2$.

Notes

1. See S.A. Hovanessian and L.A. Pipes, *Digital Computer Methods in Engineering* (New York: McGraw-Hill, 1969); and John Todd *Survey of Numerical Analysis* (New York: McGraw-Hill, 1962).

2. For another corrector formula, see Hovanessian and Pipes, *Digital Computer Methods*, p. 233.

3. C. Jordan, *Calculus of Finite Differences*, (New York: Chelsea, 1947), chap. 2.

Bibliography

Abraham, F.F., and Tiller, W.A. *An Introduction to Computer Simulation in Applied Science*. New York: Plenum Press, 1972.

Froberg, C.E. *Numerical Analysis*. Reading, Mass.: Addison-Wesley, 1965.

Hamming, R.W. *Numerical Methods for Scientists and Engineers*. New York: McGraw-Hill, 1973.

Hildebrand, F.B. *Finite Difference Equations and Simulations*. Englewood Cliffs, N.J.: Prentice-Hall, 1968.

Hovanessian, S.A., Pipes, L.A., *Digital Computer Methods in Engineering*. New York: McGraw-Hill, 1969.

James, M.L., Smith, G.M., and Wolford, J.C. *Applied Numerical Methods for Digital Computations*. Scranton, Pa.: International Textbook, 1967.

Lapidus, L., and Seinfeld, J.H. *Numerical Solution of Ordinary Differential Equations*. New York: Academic Press, 1971.

Martin, H.C., and Carey, G.F. *Introduction to Finite Element Analysis*. New York: McGraw-Hill, 1973.

Ralston, A., and Wilf, H.S. (eds.) *Mathematical Models for Digital Computers*. New York: Wiley, 1960 (vol. 1); 1967 (vol. 2). Somewhat more advanced than the other references cited here.

Von Rosenberg, Dale U. *Methods of the Numerical Solution of Partial Differential Equations*. New York: American Elsevier, 1960.

Appendix 3A
Digital Mechanization of Finite Difference Formulations

With the advent of digital components, analog systems similar to the one in Figure 3-2 may be replaced by equivalent digital systems. These digital systems use the usual arithmetical operations such as addition and multiplication and also make use of delay lines. The digital systems are derived from finite difference representations, such as those given by equations (3.79) and (3.80). By use of these equations, the digital signal-processing systems equivalent to that of Figure 3-2 can be drawn as shown in Figure 3A-1.

Note the delay lines with a time delay which is equal to the time step T shown in Figure 3-3. The input signal σ is sampled at intervals of T, starting from zero. Note also that the delay lines hold the values of γ_{n+1} and ϵ_{n+1} for the time duration T. These values are to be used in the next time-step computation, where σ_{n+1} is sampled as the input signal.

In a similar fashion, one can draw the equivalent digital circuit for Milne's predictor-corrector formulation of equations (3.38) and (3.40), as shown in Figure 3A-2. In this case, delay components are denoted by T, and f_n from equation (3.35) represents the value of derivative y' at time step n. The parameters α and β are constants representing coefficients of equations (3.38) and (3.40).

The nonrecursive portion (the predictor equation [3.38]) is shown in Figure 3A-2 with solid boundaries, while the recursive portion (corrector equation [3.40]) is shown by dashed lines. Figure 3A-2 may be considered an example of a recursive digital filter.

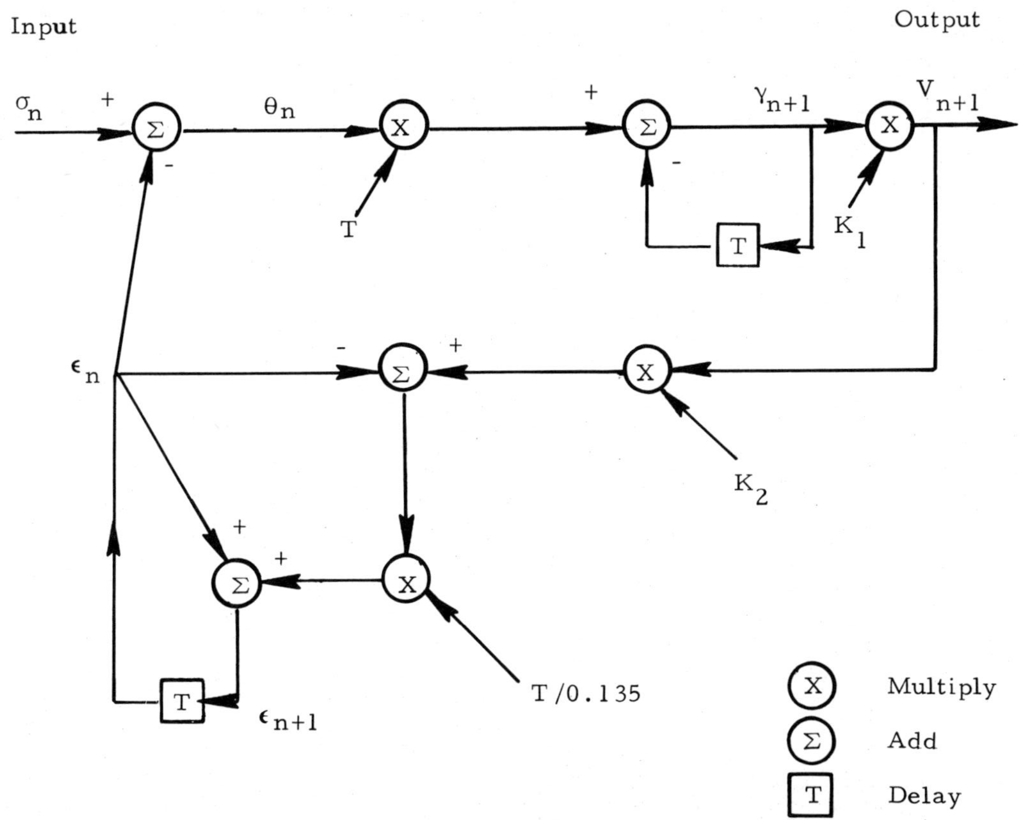

Figure 3A-1. Digital Signal-Processing Equivalent to the Analog Servo System of Figure 3-2

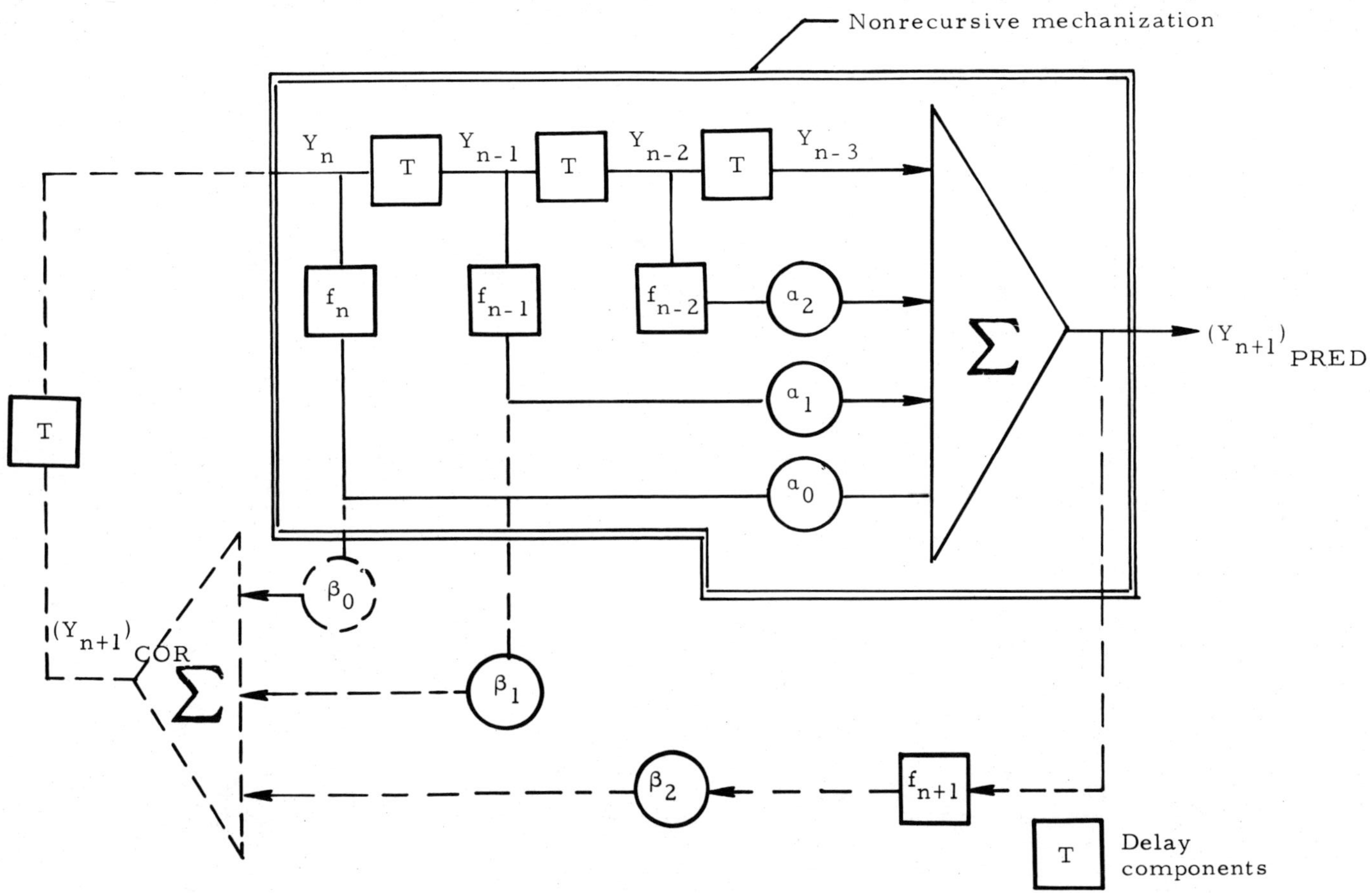

Figure 3A-2. Digital Mechanization of Predictor-Corrector Equations (3.38) and (3.40)—Recursive Digital Filter

4 Probability Theory and Monte Carlo Simulations

Within the last twenty years the methods of probability theory have been applied to a wide variety of engineering situations. The primary application of the probability theory has been under the conditions where an exact solution of the problem could not have been obtained by the use of classical rules of physics and mathematics. These conditions included cases where past data were analyzed to obtain future trends and situations dealing with random motions such as arise in noise theory.

Although classical probability theory has been in existence for over one hundred years, only recently have these theories been applied to the solution of engineering problems arising in industry. In some branches of engineering, these applications have been introduced because of reliability demands on modern complicated systems. The reliability of these systems, expressed as mean time to failure, is predicted and/or analyzed by using probability theories and appropriate probability distributions.

Actual physical situations, which include measurement errors, component accuracies, and random inputs, are incorporated in mathematical models of engineering problems through the use of probability methods. The digital computer simulations of these models with probabilistic inputs are referred to as Monte Carlo simulations.

Basic Ideas of Probability Theory

The value of a probability, P, is a number between 0 and 1, both limits included

$$0 \leq P \leq 1 \tag{4.1}$$

The probability of a certain event is 1.0

The probability of an impossible event is 0.0

The probability that at least one of the two events occurs is the sum of the probabilities of each event minus the probability of both events occurring simultaneously

$$P(A + B) = P(A) + P(B) - P(AB) \tag{4.2}$$

where A and B are events and $P(A + B)$ is the probability of either A or B occurring and $P(AB)$ is the probability of both A and B occurring simultaneously.

Example: One card is drawn from each of two decks. The probability that either one of them is the queen of hearts is

$$\begin{aligned} P &= 1/52 + 1/52 - (1/52)(1/52) \\ &= 103/2704 = 3.8 \text{ percent} \end{aligned} \tag{4.3}$$

where 52 is the number of cards in the deck.

The probability of both of two events occurring is equal to the product of the absolute probability of one event and the conditional probability of the other

$$\begin{aligned} P(AB) &= P(A)\,P(B/A) \\ &= P(B)\,P(A/B) \end{aligned} \tag{4.4}$$

Example: We draw two cards from a deck. The probability that both cards are hearts is given by

$$\frac{13}{52} \cdot \frac{12}{51} = \frac{1}{17} = 5.9 \text{ percent} \tag{4.5}$$

where 52 is the number of cards in the deck and 13 is the number of hearts. After one heart is removed, the corresponding numbers become 51 and 12.

If two events exclude each other, their joint probability, $P(AB)$, is zero. From (4.2), we have

$$P(A + B) = P(A) + P(B) \tag{4.6}$$

Two events are independent if the probability of both occurring is equal to the product of probability of each event occurring separately.

$$P(AB) = P(A)\,P(B) \tag{4.7}$$

Example: One card is drawn from each of two decks; the probability that both are hearts is then

$$\frac{13}{52} \cdot \frac{13}{52} = \frac{1}{16} = 6.25 \text{ percent} \tag{4.8}$$

The probability of event A occurring is P. The probability of event A not occurring is $(1 - P)$.

Bernoulli's Problem

Consider event A with a probability of occurring θ, $0 \leq \theta \leq 1$. We perform v observations and ask for the probability P_r that A occurs in r, $0 \leq r \leq v$ of these observations.

Solution: First assume A occurs in the first r observations and not the rest. The probability of this event will be

$$\theta^r (1 - \theta)^{v-r} \tag{4.9}$$

But A can occur r times any place in v observations. The number of ways we can select r elements in v observations is

$$\binom{v}{r} = \frac{v(v-1)\ldots(v-r+1)}{1 \cdot 2 \cdot 3 \cdot \ldots r} = \frac{v!}{r!(v-r)!} \tag{4.10}$$

Consequently

$$P_r = \binom{v}{r} \theta^r (1 - \theta)^{v-r} \tag{4.11}$$

where the symbol ! in equation (4.10) denotes factorial products. Equation (4.11) is widely used in engineering applications.

First and Second Moments

Consider variable t with the *probability density function* $\psi(t)$. The integral of the function $\psi(t)$ between the limits t_1 and t_2 defines the probability of variable t being in the interval t_1 to t_2; i.e.,

$$\int_{t_1}^{t_2} \psi(t)\, dt = P(t_1 < t \leq t_1)$$

$$= \phi(t_2) - \phi(t_1) \tag{4.12}$$

where P is the probability. The integral in (4.12) is usually used in defining the probability density. The function $\phi(a)$ is known as the probability distribution function, $\phi(a) = \int_{-\infty}^{a} \psi(t)\, dt$.

The *first moment* of $\psi(t)$ is defined as

$$M_1 = \int_{-\infty}^{\infty} t\, \psi(t)\, dt \tag{4.13}$$

Similarly, the nth moment of $\psi(t)$ is defined as

$$M_n = \int_{-\infty}^{\infty} t^n \psi(t)\, dt$$

$$n = 1, 2, 3, \ldots \tag{4.14}$$

The probability density function can be normalized by using the integral

$$N = \int_{-\infty}^{\infty} \psi(t)\, dt \tag{4.15}$$

where N is the normalization constant. The integral (4.15) defines the probability of t assuming values between $-\infty$ and ∞. Since this probability is equal to unity, we will have

$$N = \int_{-\infty}^{\infty} \psi(t)\, dt = 1 \tag{4.16}$$

where N represents the probability of t assuming values between $-\infty$ to ∞.

Dividing equation (4.14) by (4.15) we get the *mean* or the *average* value μ

$$\mu = \frac{M_1}{N} = \frac{\int_{-\infty}^{\infty} t\psi(t)\, dt}{\int_{-\infty}^{\infty} \psi(t)\, dt} = \int_{-\infty}^{\infty} t\psi(t)\, dt \tag{4.17}$$

Note that the idea of moments in probability theory parallels that of the dynamics of rigid bodies where the center of gravity is defined. In this context, equation (4.17), in effect, defines the center of gravity and equations (4.13) and (4.15) define the first moment and the mass of the rigid body.

Similarly, the second central moment about the mean μ of the probability density function $\psi(t)$, corresponding to the moment of inertia of rigid bodies, can be written

$$\begin{aligned} \bar{M}_2 &= \int_{-\infty}^{\infty} (t - \mu)^2 \psi(t)\, dt \\ &= \int_{-\infty}^{\infty} t^2\psi(f) - 2\mu \int_{-\infty}^{\infty} t\psi(t) + \mu^2 \int_{-\infty}^{\infty} \psi(t)\, dt \\ &= M_2 - \mu^2 \end{aligned} \tag{4.18}$$

where equations (4.16) and (4.17) are used in the derivation and M_2 represents the second moment of $\psi(t)$.

Normalizing with respect to equation (4.15), we get

$$\sigma^2 = \frac{\int_{-\infty}^{\infty} (t - \mu)^2\, \psi(t)\, dt}{\int_{-\infty}^{\infty} \psi(t)\, dt} \tag{4.19}$$

Using (4.16), we get

$$\sigma^2 = \int_{-\infty}^{\infty} (t - \mu)^2 \psi(t)\, dt \tag{4.20}$$

where σ^2 is the variance and σ is the *standard deviation* of the probability density function $\psi(t)$.

Discrete Case

These formulas can be extended to discrete data points using the following derivation: Consider a set of m *individual* data points $\tau_1, \tau_2, \ldots, \tau_m$. Assume that these data points are divided into n groupings or classes of data points $(n < m)$ $t_1, t_2, \ldots, t_n$ with intervals of $t_2 - t_1 = t_3 - t_2 = \ldots = \Delta t$. The *individual* data points τ are put in this grouping so that

$$t_i - \frac{\Delta t}{2} < \tau \leq t_i + \frac{\Delta t}{2} \tag{4.21}$$

The number of data points in each grouping is designated by f_i. This f_i can be related to the continuous case with the probability density function $\psi(t)$ through the integral.

$$f_i = m\int_{t_i - \Delta t/2}^{t_i + \Delta t/2} \psi(t)\,dt \tag{4.22}$$

where m is the number of data points. Note also that $\Sigma_{i=1}^{n} f_i = m$. By use of these relations, equation (4.17) can be written for the discrete case as follows:

$$\begin{aligned} \mu &= \sum_{i=1}^{n} t_i \int_{t_i - \Delta t/2}^{t_i + \Delta t/2} \psi(t)\,dt \\ &= \sum_{i=1}^{n} t_i f_i \Big/ m \end{aligned} \tag{4.23}$$

where μ is the average or mean value of data t_i with frequency of occurrence f_i. Similarly, equation (4.20) for the discrete case becomes

$$\sigma^2 = \sum_{i=1}^{n} (t_i - \mu)^2 f_i \Big/ m \tag{4.24}$$

where σ^2 is the variance and m is the number of data points. Equation (4.24) can also be put in the equivalent form

$$\sigma^2 = \left[\sum_{i=1}^{n} t_i^2 f_i \Big/ m\right] - \mu^2 \tag{4.25}$$

For computer calculations, the above form is used in preference to (4.24), as it does not require the availability of μ prior to the start of the calculations for σ^2.

Example Problem

Consider the data of Table 4-1 obtained from a set of engineering experiments. The total number of data points $m = \Sigma_{i=1}^{9} f_i = 96$. Using this and $\Sigma_{i=1}^{9} x_i f_i = 780$, we obtain the average value μ

$$\mu = 780/96 = 8.125 \tag{4.26}$$

Obtaining the summation $(x_i - \mu)^2 f_i$ from the table and using equation (4.24), we get

$$\sigma = SQRT(12862.62/96)$$

$$= 11.58 \tag{4.27}$$

Expected Values

To extend the ideas of moments discussed earlier to more than one dimension, we bring in the notion of expected values. The expected value of variable x with a probability density function $f(x)$ is defined as

$$E(x) = \int_{-\infty}^{\infty} xf(x)\,dx \tag{4.28}$$

where $E(x)$ is the expected value of x. Note that the expected value defined in (4.28) is the same as the first moment defined in equation (4.13). Now, consider the two-dimensional case of variables x and y with the joint probability density function $f(x, y)$. The equation corresponding to (4.16) in this case becomes

$$\int_{-\infty}^{\infty}\int_{-\infty}^{\infty} f(x,y)\,dxdy = 1$$

We define product moments M_{ab} as

$$M_{ab} = E(x^a, y^b) = \int_{-\infty}^{\infty}\int_{-\infty}^{\infty} x^a y^b f(x,y)\,dxdy \tag{4.29}$$

where $E(x^a y^b)$ is the expected value and a and b are nonnegative integers. For $a = b = 1$, equation (4.29) reduces to

$$M_{11} = E(xy) = \int_{-\infty}^{\infty}\int_{-\infty}^{\infty} xyf(x,y)\,dxdy \tag{4.30}$$

The *central* moment product $\bar{M}_{11}$ can be written about the mean values of variables x and y; μ_x and μ_y, respectively, as

Table 4-1
Data and Calculation for Example Problem

i	x_i	f_i	x_if_i	$(x_i - \mu)^2f_i$
1	−30	0	0	0
2	−20	2	−40	1582.03
3	−10	9	−90	2956.64
4	0	28	0	1848.44
5	10	30	300	105.47
6	20	21	420	2961.42
7	30	5	150	2392.60
8	40	1	40	1016.02
9	50	0	0	0
	Totals	96	780	12862.62

$$\bar{M}_{11} = E[(x - \mu_x)(y - \mu_y)]$$

$$= \int_{-\infty}^{\infty}\int_{-\infty}^{\infty} (x - \mu_x)(y - \mu_y) f(x,y)\, dxdy$$

$$= M_{11} - \mu_x\mu_y \tag{4.31}$$

The moment product is called the *covariance* of the variables x and y. Using this, we define the correlation coefficient ρ

$$\rho = \frac{\bar{M}_{11}}{\sigma_x\sigma_y} = \frac{M_{11} - \mu_x\mu_y}{\sigma_x\sigma_y} \tag{4.32}$$

where σ_x^2 and σ_y^2 are the variance of variables x and y.

Discrete Case

Consider the discrete data consisting of variables x and y in the form x_1, $x_2, \ldots, x_n$ and $y_1, y_2, \ldots, y_n$ with the frequency of occurrence $f_1, f_2, \ldots, f_n$. The equation, corresponding to (4.22) in this case can be written

$$f_i = m\int_{x_i-(\Delta x/2)}^{x_i+(\Delta x/2)}\int_{y_i-(\Delta y/2)}^{y_i+(\Delta y/2)} f(x,y)\, dxdy \tag{4.33}$$

where m is the number of data points and Δx and Δy are the dimensions of grouping of variables x and y. The total number of data points m is given as

$$m = \sum_{i=1}^{n} f_i \tag{4.34}$$

Using equation (4.33) in (4.30) and (4.31), we get the expected values[a] about $x = 0$, $y = 0$ and $x = \mu_x$, $y = \mu_y$, respectively as

$$M_{11} = E(xy) = \frac{1}{m} \sum_{i=1}^{n} x_i y_i f_i \tag{4.35}$$

$$\bar{M}_{11} = E[(x - \mu_x)(y - \mu_y)] = \frac{1}{m} \sum_{i=1}^{n} (x_i - \mu_x)(y_i - \mu_y) f_i \tag{4.36}$$

The correlation coefficient ρ can be written using M_{11} and M_n from (4.35) and (4.36) as

$$\rho = \sum_{i=1}^{n} x_i y_i f_i / m\sigma_x \sigma_y \tag{4.37}$$

and

$$\bar{\rho} = \sum_{i=1}^{n} (x_i - \mu_x)(y_i - \mu_y) f_i / m\sigma_x \sigma_y \tag{4.38}$$

where σ_x and σ_y are the *variance* of variables x and y. Note that for 100 percent correlation, i.e., $x_i = y_i$ for all i, the value of the correlation coefficient ρ becomes unity.

Properties of Expected Values

We note that the definition of expected value $E(x)$ given in (4.28) is exactly the same as that of the mean value defined in (4.17). Thus, the expected value of a variable x with the probability density function $f(x)$ is equal to the mean value of the variable.

Setting $x = y$ in (4.31), we see that this equation becomes the same as equation (4.20) defining the variance. Thus, the expected value $E(x - \mu_x)^2$ is equal to the variance σ^2 of the variable x with the probability density function $f(x)$. Defining

$$E(g(x)) = \int_{-\infty}^{\infty} g(x) f(x)\, dx \tag{4.39}$$

the following additional properties of expected values can easily be deduced:

$$E(cg) = CE(g)$$

$$E(g_1 + g_2) = E(g_1) + E(g_2) \tag{4.40}$$

[a] The number of data points m is sometimes dropped from the definition of expected values.

where C is a constant and g, g_1, and g_2 are any functions of a set of random variables. The second equation (4.40) comes from the fact that the integral of a sum is the sum of the integrals.

If g_1 and g_2 are independently distributed, which means that their joint probability density function can be factored in the form $f_1(g_1)f_2(g_2)$ from (4.30), we get

$$E(g_1g_2) = \int_{-\infty}^{\infty}\int_{-\infty}^{\infty} g_1g_2f_1(g_1)f_2(g_2)\,dg_1\,dg_2 \tag{4.41}$$

resulting in

$$\begin{aligned} E(g_1g_2) &= \int_{-\infty}^{\infty} g_1f_1(g_1)\,dg_1\int_{-\infty}^{\infty} g_2f_2(g_2)\,dg_2 \\ &= E(g_1)E(g_2) \end{aligned} \tag{4.42}$$

These derived probability relations will be used in the study of the least-squares methods and estimates in Chapter 5.

Example

Consider the coordinate system 1-2 and the orthogonal vectors **A** and **B** (see Figure 4-1).

$$\begin{aligned} &\mathbf{A}{:}\ x_1 = 2, \quad x_2 = 0 \\ &\mathbf{B}{:}\ y_1 = 0, \quad y_2 = 1 \end{aligned} \tag{4.43}$$

Show that the expected value $E(AB)$ and therefore the correlation coefficient ρ for this case will be zero.

From equation (4.35), we have

$$\begin{aligned} M_{11} = E(AB) &= x_1y_1 + x_2y_2 \\ &= (2)(0) + (0)(1) \\ &= 0 \end{aligned} \tag{4.44}$$

As a variation of this example, consider the expected value of vector $\mathbf{C}(z_1 = 1, z_2 = 1)$ with vector **A**; i.e., $E(\mathbf{AC})$. Using equation (4.35), we get

$$\begin{aligned} E(\mathbf{AC}) &= x_1z_1 + x_2z_2 \\ &= 2 \end{aligned} \tag{4.45}$$

Normalizing this with repsect to the length of vectors **A** and **C**, we get the correlation coefficient ρ

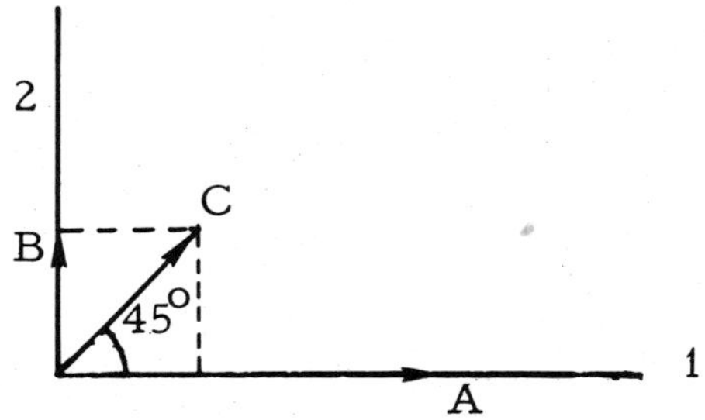

Figure 4-1. Orthogonal Vectors **A** and **B** of Equation (4.43)

$$\rho = \frac{2}{2\sqrt{(2)}} = \frac{\sqrt{(2)}}{2} \tag{4.46}$$

which is the cosine of the angle between the two vectors **A** and **C**. Thus, vectors **A** and **C** are correlated through the cosine of the subtended angle.

Normal Distribution

The probability density function of the normal or Gaussian distribution is given by

$$d\phi = \psi(t)\,dt = \frac{1}{\sigma\sqrt{(2\pi)}} \exp\left[-\frac{(t-\mu)^2}{2\sigma^2}\right] dt \tag{4.47}$$

where μ is the mean and σ is the standard deviation. We also have

$$\frac{1}{\sigma\sqrt{(2\pi)}} \int_{-\infty}^{\infty} \exp\left[-\frac{(t-\mu)^2}{2\sigma^2}\right] dt = 1 \tag{4.48}$$

The normalized distribution is obtained by setting $\mu = 0$ and $\sigma = 1$

$$d\phi = \psi(t)\,dt = \frac{1}{\sqrt{(2\pi)}} \exp(-t^2/2)\,dt \text{ (normalized)} \tag{4.49}$$

The cumulative distribution is obtained as an integral of equation (4.49).

$$\phi(t) = \frac{1}{\sqrt{(2\pi)}} \int_{-\infty}^{t} \exp(-t^2/2)\,dt$$

$$\text{(tabulated in probability texts)} \tag{4.50}$$

Graphs of $\phi(t)$ and $\psi(t)$ are shown in Figure 4-2. The normalized cumulative distribution curve $\phi(t)$ starts from zero, theoretically at $t = -\infty$, and

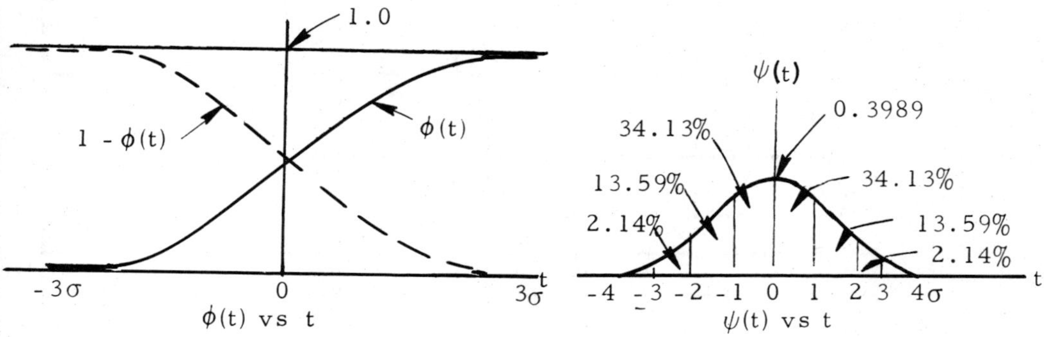

Figure 4-2. Graphs of Normal Distribution

increases to one at $t = +\infty$. For practical purposes, however, the range of values is between $-3\sigma \leq t \leq 3\sigma$. Note that 50 percent of the values are above $t = 0$ and 50 percent are below $t = 0$.

The probability density $\psi(t)$ versus t is also shown in Figure 4-2. This is the well-known bell-shaped curve of the Gaussian or normal distribution. This curve is symmetrical with respect to $\psi(t)$ axis and it shows that 68.26 percent $\simeq$ 68 percent of the values are between the limits of plus and minus one sigma.

In a number of engineering calculations one may utilize the normal distribution probability paper shown in Figure 4-3. Note that in this figure the mean value μ occurs at 50 percent and $\mu \pm \sigma$ occurs at 50 ± 34 percent, respectively. The probability values derived from equations given in this section are tabulated in almost every probability book.[1] A set of useful tables dealing with the probability distributions in this chapter is given in Appendix 4A.

Dependence on the Number of Data Points

The dispersion of the mean and sigma values calculated from measurements $x_1, x_2, \ldots, x_n$ can be represented as

$$\mu \cong \bar{X} \quad \text{with dispersion} \quad \sigma(\bar{X}) = \frac{\sigma}{\sqrt{(n)}} \tag{4.51}$$

and

$$\sigma \cong s \quad \text{with dispersion} \quad \sigma(s) = \frac{s}{\sqrt{[2(n-1)]}} \tag{4.52}$$

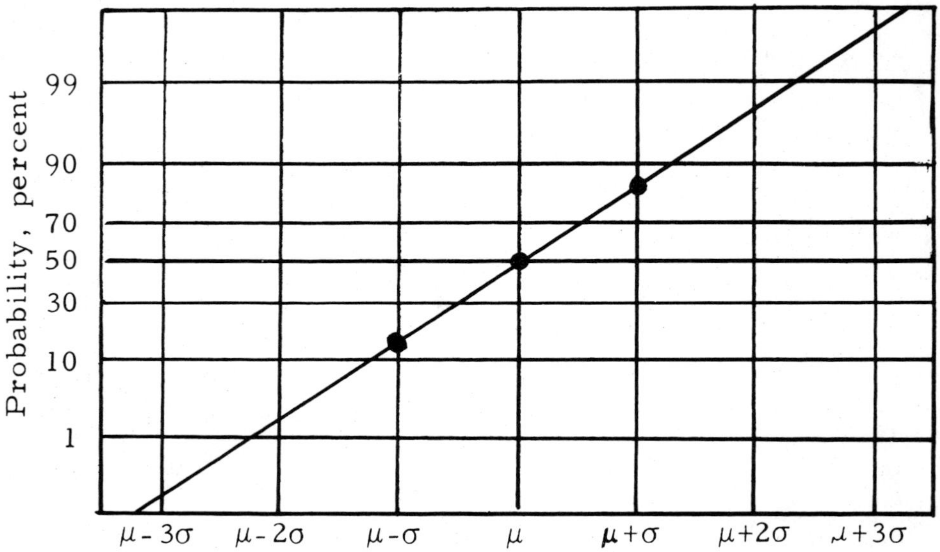

Figure 4-3. Gaussian Probability Paper

where n is the number of data points. From equations (4.51) and (4.52), it is seen that a large number of data points n reduces the dispersions on the calculated mean and sigma values.

Functions of Variables

Consider the function $Z = f(x_1, x_2, \ldots, x_n)$ of independent variables $x_1, \ldots, x_n$. Approximations to the mean and σ values of Z can be obtained by a consideration of the mean and sigma values of the independent variables $x_1, \ldots, x_n$. The appropriate equations are as follows:

$$m(z) = \bar{Z} = f(\bar{x}_1, \ldots, \bar{x}_n)$$

$$\sigma^2(z) = s^2 = \left(\frac{\partial f}{\partial x_1}\right)^2 s_1^2 + \ldots + \left(\frac{\partial f}{\partial x_n}\right)^2 s_n^2 \qquad (4.53)$$

where $x_1, \ldots, x_n$ and $s_1, \ldots, s_n$ represent the mean and the sigma values of independent variables $x_1, \ldots, x_n$.[2]

Exponential Distribution

The following represents the probability density function of an exponential distribution,

$$d\phi = \psi(t)\,dt = \frac{1}{\mu}\exp(-t/\mu)\,dt \qquad (4.54)$$

where t is the variable and μ is the mean value. The normalized exponential distribution ($\mu = 1$) is given

$$d\phi = \psi(t)\,dt = \exp(-t)\,dt \qquad (4.55)$$

with the cumulative distribution

$$\phi(t) = \int_0^t \exp(-t)\,dt = -e^{-t}\Big|_0^t = 1 - e^{-t} \qquad (4.56)$$

Graphs of $\phi(t)$ and $\psi(t)$ are given in Figure 4-4. From the graph of $\phi(t)$ versus t, it can be seen that 63 percent of the values are below $t = 1.0$. This means that in a set of data points following an exponential distribution, 63 percent of these data points will be below (37 percent above) the mean value μ.

Exponential distribution of equation (4.56) will result in a straight line on *semilogarithmic* paper as seen below

$$1 - \phi(t) = e^{-t} \quad \text{and} \quad 1 - \phi(t) = \log e^{-t} = -t \log e \qquad (4.57)$$

The last equation on a logarithmic scale is the same as $y = ax$, which is the equation of a straight line.

Goodness-of-Fit Tests

To answer the question of how well a set of observations fits a given distribution, we may use the χ^2 tests. In this test, we compute the χ^2 function (chi-squared function)

$$\chi^2 = \sum_{i=1}^{k}\frac{(f_{oi} - f_{ei})^2}{f_{ei}} \qquad (4.58)$$

where f_{oi} = observed frequency in each group or class

f_{ei} = expected (theoretical) frequency in each group or class

k = number of groups or classes (related to degrees of freedom)

The values of χ^2 are tabulated in probability books as a function of degrees of freedom.

The degrees of freedom, in turn, are related to the number of constraints put on them. For example, if we have n data points $x_1, x_2, x_3, \ldots, x_n$ and also use the average (or mean) value $\bar{x}$, we will have $n - 1$ degrees of freedom. The reason is that, having the average value and $(n - 1)$ data points, we can compute the nth data point. The assumption of a particular distribution may also be considered as a constraint on the data, reducing the number of degrees of freedom.

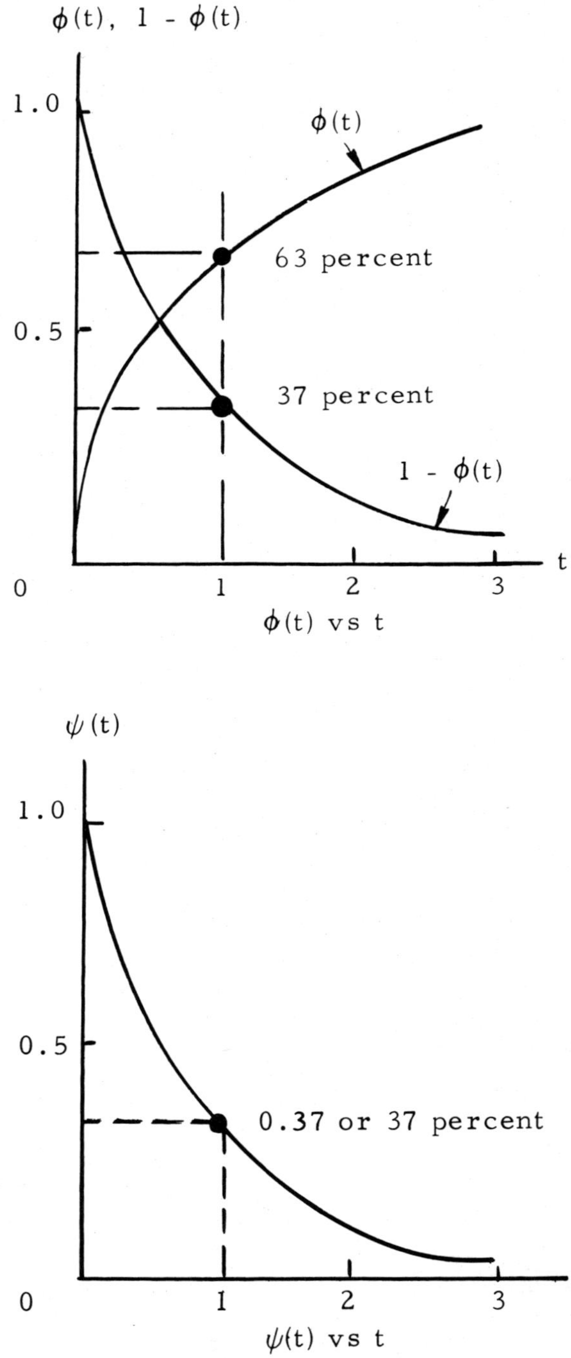

Figure 4-4. Graphs of Exponential Distribution

Example: Assume the following frequency values are obtained from 150 rolls of one die. One through 6 appeared 33, 23, 24, 23, 28, and 19 times, respectively. The question is: Is this reasonable?

Solution: Theoretically, numbers 1 through 6 should have appeared 25 times each in 150 rolls, assuming a uniform distribution. Using this, the χ^2 function can be written from (4.58) as

$$\chi^2 = \frac{(33 - 25)^2}{25} + \frac{(23 - 25)^2}{25} + \ldots + \frac{(19 - 25)^2}{25} = 4.72 \qquad (4.59)$$

There were six groups of collected data, resulting—with the assumption of uniform distribution—in 5 degrees of freedom (6 − 1). From the table of χ^2 distribution, we find that for 5 degrees of freedom there is a

$$59 \text{ percent chance of } \chi^2 > 4.351$$

$$30 \text{ percent chance of } \chi^2 > 6.064 \qquad (4.60)$$

Interpolating, we see that there is a 45 percent chance of obtaining a χ^2 as large as 4.72. We may assume, therefore, that the die is *unbiased* (unloaded) and the occurrence is purely by chance.

Kolmogorov-Smirnov Test

Another statistical test that is used to compute the degree of goodness of fit of experimental data or computer results to a given distribution is the Kolmogorov-Smirnov (K and S) test. Because of the availability of tables, this test is a somewhat easier to apply than the χ^2 test. The K and S table of goodness-of-fit test is given in Table 4-2.

The K and S goodness-of-fit significance levels given in Table 4-2 are primarily for cases where a statistical sample is tested against a completely specified distribution with defined parameters (such as the mean and standard deviation in the case of Gaussian distribution). If the parameters are not defined but are calculated for the statistical sample, which is usually the case in engineering applications, Table 4-2 yields a somewhat optimistic probability of goodness-of-fit value. H.W. Lilliefors presents two sets of tables—one for Gaussian and one for exponential distribution—where it is assumed that at specified statistical distribution parameters are obtained from a measured sample.[3] The application of the K and S method is best explained by numerical examples.

Example 1: Consider the data points obtained for mean time between failures (MTBFs) of a large electronic system, as given in Table 4-3. A plot of data points against an exponential distribution is also given in Figure 4-5.

Table 4-2
Kolmogorov-Smirnov Numerics for Goodness-of-Fit Test

Sample size	Significance level (probability)											
	0.10[a] (0.90)	0.20 (0.80)	0.30 (0.70)	0.40 (0.60)	0.50 (0.50)	0.60 (0.40)	0.70 (0.30)	0.80 (0.20)	0.90 (0.10)	0.95 (0.05)	0.98 (0.02)	0.99 (0.01)
1	0.55	0.60	0.65	0.70	0.75	0.80	0.85	0.90	0.95	0.975	0.99	0.995
2	0.40	0.43	0.46	0.49	0.53	0.57	0.62	0.68	0.78	0.84	0.90	0.93
3	0.33	0.35	0.38	0.41	0.44	0.47	0.51	0.56	0.64	0.71	0.79	0.83
4	0.29	0.31	0.33	0.36	0.38	0.41	0.45	0.49	0.56	0.62	0.69	0.73
5	0.26	0.28	0.30	0.32	0.34	0.37	0.41	0.45	0.51	0.56	0.63	0.67
6	0.24	0.25	0.27	0.29	0.31	0.34	0.37	0.41	0.47	0.52	0.58	0.62
7	0.22	0.24	0.25	0.27	0.29	0.32	0.34	0.38	0.44	0.49	0.54	0.58
8	0.21	0.22	0.24	0.26	0.28	0.30	0.32	0.36	0.41	0.46	0.51	0.54
9	0.20	0.21	0.23	0.24	0.26	0.28	0.31	0.34	0.39	0.43	0.48	0.51
10	0.19	0.20	0.22	0.23	0.25	0.27	0.29	0.32	0.37	0.41	0.46	0.49
11	0.18	0.19	0.21	0.22	0.24	0.26	0.28	0.31	0.35	0.39	0.44	0.47
12	0.17	0.19	0.20	0.21	0.23	0.25	0.27	0.29	0.34	0.38	0.42	0.45
13	0.16	0.18	0.19	0.21	0.22	0.24	0.26	0.28	0.32	0.36	0.40	0.43
14	0.16	0.17	0.18	0.20	0.21	0.23	0.25	0.27	0.31	0.35	0.39	0.42
15	0.16	0.17	0.18	0.19	0.21	0.22	0.24	0.27	0.30	0.34	0.38	0.40
16	0.15	0.16	0.17	0.19	0.20	0.22	0.23	0.26	0.30	0.33	0.37	0.39
17	0.15	0.16	0.17	0.18	0.19	0.21	0.23	0.25	0.29	0.32	0.36	0.38
18	0.14	0.15	0.16	0.18	0.19	0.20	0.22	0.24	0.28	0.31	0.35	0.37
19	0.14	0.15	0.16	0.17	0.18	0.20	0.22	0.24	0.27	0.30	0.34	0.36
20	0.14	0.15	0.16	0.17	0.18	0.19	0.21	0.23	0.26	0.29	0.33	0.35
25	0.12	0.13	0.14	0.15	0.16	0.17	0.19	0.21	0.24	0.26	0.30	0.32
30	0.11	0.12	0.13	0.14	0.15	0.16	0.17	0.19	0.22	0.24	0.27	0.29
35	0.10	0.11	0.12	0.13	0.14	0.15	0.16	0.18	0.21	0.23	0.25	0.27
> 35	$\frac{0.61}{\sqrt{n}}$	$\frac{0.65}{\sqrt{n}}$	$\frac{0.70}{\sqrt{n}}$	$\frac{0.75}{\sqrt{n}}$	$\frac{0.81}{\sqrt{n}}$	$\frac{0.87}{\sqrt{n}}$	$\frac{0.95}{\sqrt{n}}$	$\frac{1.07}{\sqrt{n}}$	$\frac{1.22}{\sqrt{n}}$	$\frac{1.36}{\sqrt{n}}$	$\frac{1.48}{\sqrt{n}}$	$\frac{1.63}{\sqrt{n}}$

[a] Column headings not in parentheses give probabilities of maximum absolute differences between observed cumulative frequencies and theoretical cumulative frequencies being smaller than or equal to the table values. The probabilities of maximum absolute differences exceeding the table values are in parentheses.

Source: B.L. Amstadter, *Reliability Mathematics* (New York: McGraw-Hill), 1971, p. 78. Reprinted with permission.

Since the data consist of 10 data points, each point has a weight of 10 percent. Using this, we compute the cumulative distribution of the data as given in column 2. Column 3 is computed using the cumulative exponential distribution probability with $\bar{x} = 11.64$. Column 4 gives the absolute value of the difference between columns 2 and 3. This difference will be used in the K and S test.

Table 4-3
Data Points for Mean Time Between Failures

(1) MTBFs Hours Measured Data, x	*(2)* Cumulative Probability	*(3)* Theoretical Cumulative Prob. $e^{-x/\bar{x}}$	*(4)* Absolute Value (Col 2 Minus Col. 3)
3.2 or more	1.00	0.7596	**0.24**
4.0 or more	0.9	0.7092	0.19
4.6 or more	0.8	0.6736	0.13
8.7 or more	0.7	0.4736	0.23
10.4 or more	0.6	0.4092	0.19
13.6 or more	0.5	0.3109	0.19
14.0 or more	0.4	0.3004	0.10
15.5 or more	0.3	0.2641	0.04
20.4 or more	0.2	0.1733	0.03
22.0 or more	0.1	0.1511	0.05

$\sum = 116.4$

Average $= \bar{X} = 11.64$ hours

The largest value appearing in column 4 is 0.24. From K and S tables, the probability is greater than 50 percent that a number larger than 0.24 will be encountered for 10 data points, by chance alone. Since a probability of greater than 20 percent is considered sufficient to show that a given distribution fits the data, the assumption of the exponential distribution of the data is valid.

Example 2: Consider a Monte Carlo computer run resulting in the data of Table 4-4. The Monte Carlo run consisted of 100 computer runs resulting in 100 parameter values. These parameter values were grouped in increments of four units as given in Table 4 -4. The values of $\bar{X} = 16.99$ and $S = 9.01$ were computed using 100 data points from the Monte Carlo runs.

The first column in Table 4-4 gives the parameter groups. The second column is obtained from the output of the Monte Carlo runs. The third column is computed from the assumption of a normal or Gaussian distribution; for example, the theoretical cumulative frequency for *12 or more* is

$$12 = 16.99 - x(9.01) \tag{4.61}$$

where x is the number of standard deviations. From this, we get $x = 0.55$. From probability tables, this value corresponds to a cumulative probability of 0.71, as given in column 3. Column 4 gives the difference between Monte Carlo and theoretical Gaussian distribution values.

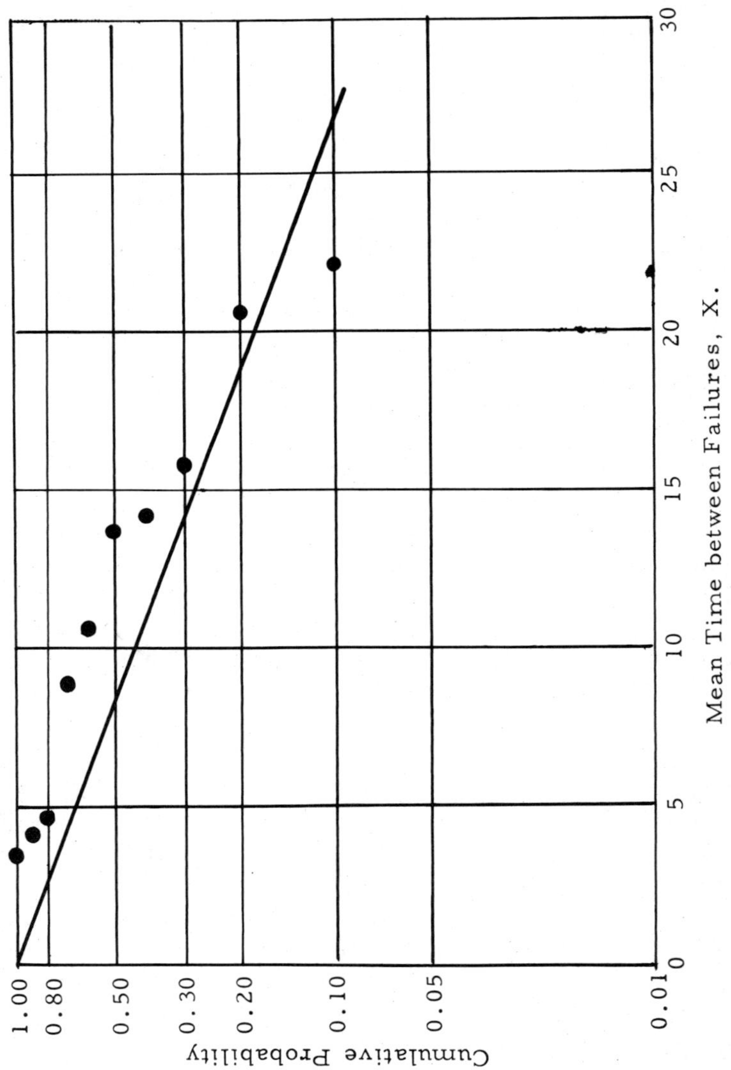

Figure 4-5. Exponential Distribution on Semilogarithmic Paper

The largest difference appearing in Table 4-4, column 4, is 0.10. Having this in mind, we go to K and S tables (Table 4-2) and get a corresponding value of 0.107, for a 20 percent probability level. This means that the probability is better than 20 percent that the 0.10 value of column 4 is due to chance and the assumption of Gaussian distribution of the Monte Carlo results is reasonable.

Tolerance Limits

Consider M data points $x_1, \ldots, x_M$ for which $\bar{X}$ and S are obtained. The limit within which x_{M+1}, a measurement yet to be made, lies with a given probability $(1 - P)$ is called the *tolerance limit*, and is obtained using t-distribution. This is true for probability distributions which are not highly peaked or skewed and also $M > 10$.

The t-distribution is defined as a function of two parameters $t(P, f)$, where P is the probability and f (is equal to $M - 1$) is the degrees of freedom. Define a new parameter.

$$\tau = \frac{\bar{X} - x_{M+1}}{S\sqrt{(1 + 1/M)}} \tag{4.62}$$

where $\bar{X}$ is the mean value, S is the standard deviation, and M is the number of data points. Equating τ is the t-distribution, we get

$$-t(P,f) \leq \tau \leq t(P,f) \tag{4.63}$$

and obtain the tolerance limits on the yet-to-be-measured values of x_{M+1} as

$$\bar{X} - S\sqrt{1 + \frac{1}{M}}\,t(P,f) \leq x_{M+1} \leq \bar{X} + S\sqrt{1 + \frac{1}{M}}\,t(P,f) \tag{4.64}$$

In practice, usually 95 or 90 percent (P = 5 or 10 percent) tolerance limits are selected.

Example: Calculate 90 percent tolerance limits of $M = 100$, $\bar{X} = 16.99$, and $S = 9.01$. We get P = 10 percent and $f = 100 - 1 = 99$. From t-distribution tables t(10 percent, 99) = 1.660 and equation (4.64) will result

$$1.96 \leq x_{M+1} \leq 32.02 \tag{4.65}$$

From this, the probability is 90 percent that the next measurement will lie between 1.96 and 32.02.

Table 4-4
Kolmogorov-Smirnov Test Applied to Monte Carlo Results

(1)	(2)	(3)	(4)
Value of Parameter	Cumulative Frequency (Monte Carlo run)	Theoretical Cumulative Frequency (Normal distribution)	Absolute Difference Column (2) – (3)
4 or more	1.00	0.93	0.07
8 or more	0.79	0.84	0.05
12 or more	0.61	0.71	0.10
16 or more	0.48	0.54	0.06
20 or more	0.32	0.37	0.05
24 or more	0.25	0.22	0.03
28 or more	0.17	0.11	0.06
32 or more	0.11	0.05	0.06
36 or more	0.00	0.02	0.02
Calculated mean $\overline{X}$ = 16.99, standard deviation S = 9.01 Number of Monte Carlo runs = 100			

Confidence Limits

Reasoning similar to the above can be used to calculate the limits of the average value $\bar{X}$, called confidence limits. These derivations apply to the cases where the distribution is not highly peaked or skewed.

In this case, we define

$$\tau = \frac{\bar{X} - \mu}{S/\sqrt{M}} \tag{4.66}$$

where τ has a t-distribution with $f = M - 1$ and μ is the true estimate of $\bar{X}$. Using τ from (4.66) in (4.63), we get the confidence limits on the value of $\bar{X}$

$$\bar{X} - \frac{S}{\sqrt{M}} t(P,f) \leq \mu \leq \bar{X} + \frac{S}{\sqrt{M}} t(P,f) \tag{4.67}$$

For the previously given example, we get the 90 percent confidence level as

$$15.494 \leq \mu \leq 18.485 \tag{4.68}$$

which means that we are 90 percent confident that our mean or average value lies between 15.494 and 18.485.

Confidence Limits (Exponential Distribution)

Confidence limits for exponential distribution with M samples are obtained by using the χ^2 distribution instead of the t-distribution. These equations are as follows:

$$\frac{2M\bar{X}}{\chi^2(\alpha/2,\ 2M)} \leq \mu \leq \frac{2M\bar{X}}{\chi^2(1 - \alpha/2,\ 2M)} \tag{4.69}$$

where $\alpha = 1 - P$, with P the probability value.

Example: For $M = 10$, $\bar{X} = 1000$, find 90 percent and 50 percent confidence limits on $\bar{X}$. From χ^2 tables, we get the 90 percent confidence level (637, 1843) and the 50-percent confidence level (839, 1294).

Generation of Random Numbers

Monte Carlo simulations require the generation of a great many random numbers. Various distributions can be obtained by using random numbers generated from a uniform distribution between zero to one.

One method of generating random numbers may be the *midsquare* method

$$
\begin{aligned}
Z_0 &= 1234 \quad Z_0^2 = 01\,\mathbf{5227}\,56 \\
Z_1 &= 5227 \quad Z_1^2 = 27\,\mathbf{3215}\,29 \\
Z_2 &= 3215 \quad Z_2^2 = 10\,\mathbf{3362}\,25 \\
Z_3 &= 3362 \quad Z_3^2 = 11\,\mathbf{3030}\,44
\end{aligned} \tag{4.70}
$$

The procedure stops, however, if **0000** occurs.

Another method often used is

$$x_i = (\pi + x_{i-1})^8 - INT(\pi + x_{i-1})^8 \tag{4.71}$$

where $0 \leq x_i \leq 1$ is the generated random number from a uniform distribution and *INT* refers to the integer portion.

Using the random numbers generated, we can generate *normal* distribution with mean and sigma values (μ, σ) from the equation

$$v = \left[\lim_{k \to \infty} \frac{\sum_{i=1}^{k} x_i - k/2}{\sqrt{(k/12)}} \right] \sigma + \mu \tag{4.72}$$

where x_i is the random number (0 to 1) generated from a uniform distribution and v is the value of the random number from a normal distribution.

The formula is particularly attractive with $k = 12$,

$$v = \left(\sum_{i=1}^{12} x_i - 6 \right) \sigma + \mu \tag{4.73}$$

Equation (4.73) is usually used to generate random numbers with a Gaussian distribution in conjunction with Monte Carlo simulations.

Monte Carlo Simulations

Monte Carlo simulations consist of several (on the order of 100) computer runs of a given system where *noise* in the form of probalistic inputs is added to various parameters at each step of iteration Δt.

For example, consider the simulation of the servo system of Figure 4-6. Assume σ has an error with a mean of $\bar{X}$ and standard deviation of S. At each iteration, the value of σ will be augmented by a random variable generated, using equation (4.73). Thus, the computed value of output V will also be augmented by random noise coming through the loop because of the noisy input. Figure 4-7 gives an example of input and output conditions for a Monte Carlo simulation.

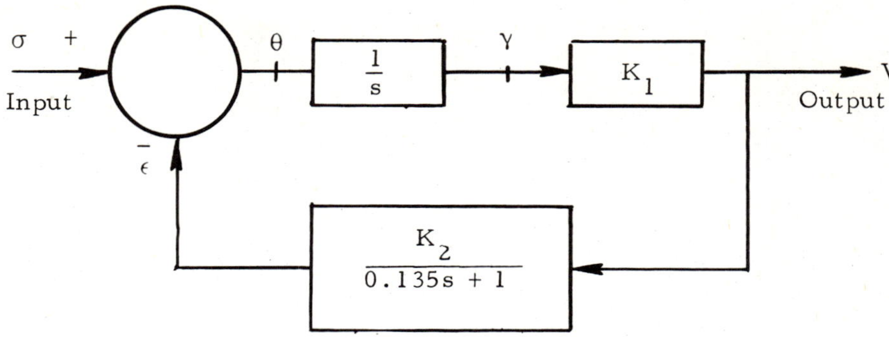

Figure 4-6. Servo System Monte Carlo Simulation

In addition to the noise added to the input σ, system noise simulating the actual workings of the circuits can also be added to the Monte Carlo simulation. For example, the integrator $1/s$ of Figure 4-6 may introduce noise into the system. This can be incorporated by adding equivalent random inputs to the integrator output γ. Other effects, such as limits on the values of specific inputs or outputs, can also be easily included in Monte Carlo simulations. Note that the values of output V are calculated at each iteration step Δt, starting with Δt and continuing to $2\Delta t$, $3\Delta t$, . . ., and so on.

For a complete Monte Carlo simulation, the problem should be run for about 100 times starting from the initial σ value. This way, at each time step Δt statistical analysis can be run on the values of calculated output V.

This, however, is not practical as it involves large computer times. Usually only one or two runs are made and behavior analyzed by plotting V versus time as seen in Figure 4-7.

Problems

1. Assuming that the probability of occurrence of an event in each try is 0.30, calculate the probability that the event occurs at least once in three tries.
2. Assuming that the probability of occurrence of an event is 0.25, calculate the probability of the event occurring three times in a row.
3. Assume that the probability of occurrence of event A is 0.20 in each experiment. We perform five experiments. Calculate the probability of A occurring in four of these five experiments.

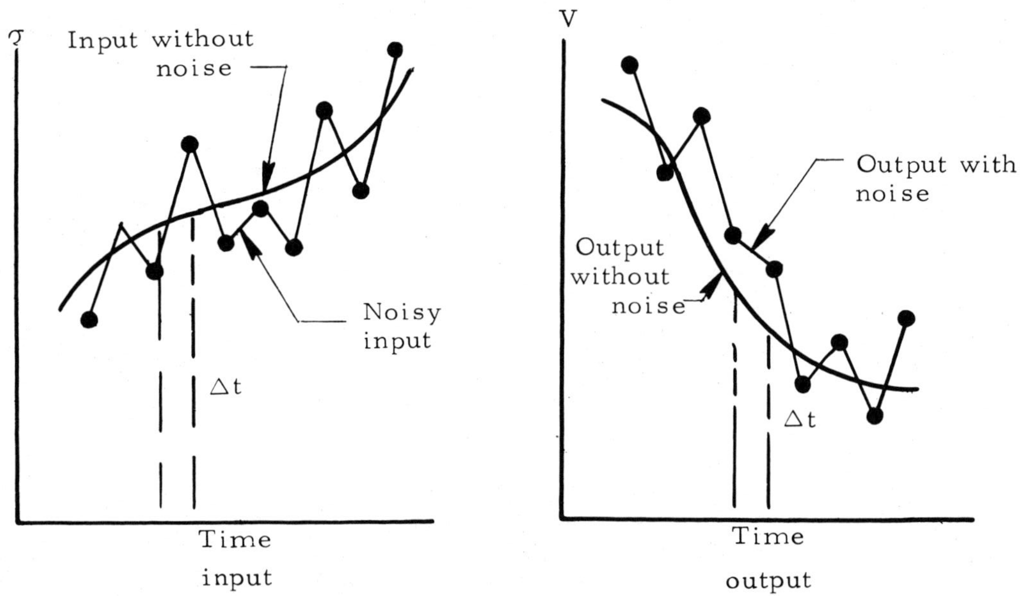

Figure 4-7. Effect of Input Noise on the Output

4. Assume that the probability of occurrence of event A is 0.3 in each experiment. We perform ten experiments. Calculate the probability of A occurring in three and five of those experiments.

5. Consider the following data:

Number of Samples	*Value*
3	50
5	60
13	70
18	80
20	90
24	100
15	110
10	120
7	130
5	140

Calculate the mean value and the standard deviation of these data. Draw the values on normal distribution paper and see if the data follow this distribution.

6. Consider the following data:

Number of Samples	*Value*
20	3
23	4
26	5
20	6
18	7
15	8
12	9
10	10
8	12
6	15
3	20
1	35

Obtain the mean value and the standard deviation of these data. Plot the data on semilogarithmic paper and see if the data follow an exponential distribution.

7. Consider the following data points:

Number of Samples	*Value*
9	2.0
8	3.0
7	3.5
6	4.5
5	6.0
4	8.0
3	11.0
2	16.0
1	24.0
1	32.0

   Obtain the mean value and the standard deviation of the above data. By plotting the data on normal distribution and semilog-arithmic paper, determine the distribution that the data follow.
8. By applying the χ^2 test to the data of problem 5, determine the degree of goodness-of-fit of these data to a Gaussian distribution.
9. By applying the χ^2 test to the data of problem 6, determine the degree of goodness-of-fit of these data to an exponential distribution.
10. Consider the set of ten data points, each occurring only once:

   105
   110
   112
   115
   120
   123
   128
   132
   133
   138

   Obtain the mean value and standard deviation of the data. Applying the Kalmogorov-Smirnov test, determine the degree of goodness-of-fit to a Gaussian distribution.
11. Apply the Kalmogorov-Smirnov test to the data of problem 5 and determine the data's goodness-of-fit to a Gaussian distribution.
12. Apply the Kalmogorov-Smirnov test to the data of problem 6 and determine the data's goodness-of-fit to the exponential distribution.
13. By applying the χ^2 test to the data

Observed frequency	15	7	4	11	6	17
Theoretical frequency	10	10	10	10	10	10

   determine the degree of goodness-of-fit.

14. Calculate the 90 percent and 80 percent tolerance limits for the data of problem 5.
15. Calculate the 90 percent and 50 percent confidence levels for the data of problem 5.
16. Derive relation

$$\sigma = \sqrt{\frac{\sum_{i=1}^{n} X_i^2}{n} - \bar{X}^2}$$

17. Consider a set of data consisting of 15 points and obeying an exponential distribution. For an average value of 87.5, compute the 90 percent and 80 percent confidence levels.
18. Consider a set of 20 data points with exponential distribution. For an average value of 1300, compute the 90 percent and 60 percent confidence levels.
19. For large values of degrees of freedom f, the variable $\sqrt{(2\chi^2)}$ is approximately normally distributed, with mean value $\sqrt{(2f - 1)}$ and dispersion $\sigma = 1$. Using this and the normal distribution tables, compute $\chi^2(90,200)$.
20. Use the method of problem 19 to compute $\chi^2(70, 145)$; $\chi^2(30, 145)$.

Notes

1. P.G. Hoel, *Mathematical Statistics* (New York: Wiley, 1966). Includes an extensive set of tables.

2. See N. Arley and K.R. Buch, *Probability and Statistics* (New York: Wiley, 1961), which is a general reference text.

3. H.W. Lilliefors, "On the Kolmogorov-Smirnov Test for Normality with Mean and Variance Unknown," *American Statistical Association Journal* (1967), p. 399; and H.W. Lilliefors, "On the Kolmogorov-Smirnov Test for the Exponential Distribution with Mean Unknown," *American Statistical Journal* (1969), p. 386.

Bibliography

Amstadter, B.L., *Reliability Mathematics*. New York: McGraw-Hill, 1971.

Fröberg, C.E., *Numerical Analysis*. Reading, Mass.: Addison-Wesley, 1965.

Hovanessian, S.A., and McRee, B.R., "Computer Analysis of Reliability

Testing," *International Journal of Computers and Electrical Engineering*, vol. 1, 1973, pp. 227-237.

Mihram, G.A., *Simulation, Statistical Foundation and Methodology*. New York: Academic Press, 1972.

Papoulis, A., *Probability, Random Variables, and Stochastic Processes*, New York: McGraw-Hill, 1965.

Appendix 4A

Table 4A-1
Normal Areas and Ordinates

t	$\phi(t)$	$\int_0^t \phi(t)dt$	t	$\phi(t)$	$\int_0^t \phi(t)dt$	t	$\phi(t)$	$\int_0^t \phi(t)dt$
.00	.39894	.00000	.45	.36053	.17364	.90	.26609	.31594
.01	.39892	.00399	.46	.35889	.17724	.91	.26369	.31859
.02	.39886	.00798	.47	.35723	.18082	.92	.26129	.32121
.03	.39876	.01197	.48	.35553	.18439	.93	.25888	.32381
.04	.39862	.01595	.49	.35381	.18793	.94	.25647	.32639
.05	.39844	.01994	.50	.35207	.19146	.95	.25406	.32894
.06	.39822	.02392	.51	.35029	.19497	.96	.25164	.33147
.07	.39797	.02790	.52	.34849	.19847	.97	.24923	.33398
.08	.39767	.03188	.53	.34667	.20194	.98	.24681	.33646
.09	.39733	.03586	.54	.34482	.20540	.99	.24439	.33891
.10	.39695	.03983	.55	.34294	.20884	1.00	.24197	.34134
.11	.39654	.04380	.56	.34105	.21226	1.01	.23955	.34375
.12	.39608	.04776	.57	.33912	.21566	1.02	.23713	.34614
.13	.39559	.05172	.58	.33718	.21904	1.03	.23471	.34850
.14	.39505	.05567	.59	.33521	.22240	1.04	.23230	.35083
.15	.39448	.05962	.60	.33322	.22575	1.05	.22988	.35314
.16	.39387	.06356	.61	.33121	.22907	1.06	.22747	.35543
.17	.39322	.06749	.62	.32918	.23237	1.07	.22506	.35769
.18	.39253	.07142	.63	.32713	.23565	1.08	.22265	.35993
.19	.39181	.07535	.64	.32506	.23891	1.09	.22025	.36214
.20	.39104	.07926	.65	.32297	.24215	1.10	.21785	.36433
.21	.39024	.08317	.66	.32086	.24537	1.11	.21546	.36650
.22	.38940	.08706	.67	.31874	.24857	1.12	.21307	.36864
.23	.38853	.09095	.68	.31659	.25175	1.13	.21069	.37076
.24	.38762	.09483	.69	.31443	.25490	1.14	.20831	.37286
.25	.38667	.09871	.70	.31225	.25804	1.15	.20594	.37493
.26	.38568	.10257	.71	.31006	.26115	1.16	.20357	.37698
.27	.38466	.10642	.72	.30785	.26424	1.17	.20121	.37900
.28	.38361	.11026	.73	.30563	.26730	1.18	.19886	.38100
.29	.38251	.11409	.74	.30339	.27035	1.19	.19652	.38298
.30	.38139	.11791	.75	.30114	.27337	1.20	.19419	.38493
.31	.38023	.12172	.76	.29887	.27637	1.21	.19186	.38686
.32	.37903	.12552	.77	.29659	.27935	1.22	.18954	.38877
.33	.37780	.12930	.78	.29431	.28230	1.23	.18724	.39065
.34	.37654	.13307	.79	.29200	.28524	1.24	.18494	.39251
.35	.37524	.13683	.80	.28969	.28814	1.25	.18265	.39435
.36	.37391	.14058	.81	.28737	.29103	1.26	.18037	.39617
.37	.37255	.14431	.82	.28504	.29389	1.27	.17810	.39796
.38	.37115	.14803	.83	.28269	.29673	1.28	.17585	.39973
.39	.36973	.15173	.84	.28034	.29955	1.29	.17360	.40147
.40	.36827	.15542	.85	.27798	.30234	1.30	.17137	.40320
.41	.36678	.15910	.86	.27562	.30511	1.31	.16915	.40490
.42	.36526	.16276	.87	.27324	.30785	1.32	.16694	.40658
.43	.36371	.16640	.88	.27086	.31057	1.33	.16474	.40824
.44	.36213	.17003	.89	.26848	.31327	1.34	.16256	.40988

Table 4A-1 (continued)

t	$\phi(t)$	$\int_0^t \phi(t)dt$	t	$\phi(t)$	$\int_0^t \phi(t)dt$	t	$\phi(t)$	$\int_0^t \phi(t)dt$
1.35	.16038	.41149	1.80	.07895	.46407	2.25	.03174	.48778
1.36	.15822	.41309	1.81	.07754	.46485	2.26	.03103	.48809
1.37	.15608	.41466	1.82	.07614	.46562	2.27	.03034	.48840
1.38	.15395	.41621	1.83	.07477	.46638	2.28	.02965	.48870
1.39	.15183	.41774	1.84	.07341	.46712	2.29	.02898	.48899
1.40	.14973	.41924	1.85	.07206	.46784	2.30	.02833	.48928
1.41	.14764	.42073	1.86	.07074	.46856	2.31	.02768	.48956
1.42	.14556	.42220	1.87	.06943	.46926	2.32	.02705	.48983
1.43	.14350	.42364	1.88	.06814	.46995	2.33	.02643	.49010
1.44	.14146	.42507	1.89	.06687	.47062	2.34	.02582	.49036
1.45	.13943	.42647	1.90	.06562	.47128	2.35	.02522	.49061
1.46	.13742	.42786	1.91	.06439	.47193	2.36	.02463	.49086
1.47	.13542	.42922	1.92	.06316	.47257	2.37	.02406	.49111
1.48	.13344	.43056	1.93	.06195	.47320	2.38	.02349	.49134
1.49	.13147	.43189	1.94	.06077	.47381	2.39	.02294	.49158
1.50	.12952	.43319	1.95	.05959	.47441	2.40	.02239	.49180
1.51	.12758	.43448	1.96	.05844	.47500	2.41	.02186	.49202
1.52	.12566	.43574	1.97	.05730	.47558	2.42	.02134	.49224
1.53	.12376	.43699	1.98	.05618	.47615	2.43	.02083	.49245
1.54	.12188	.43822	1.99	.05508	.47670	2.44	.02033	.49266
1.55	.12001	.43943	2.00	.05399	.47725	2.45	.01984	.49286
1.56	.11816	.44062	2.01	.05292	.47778	2.46	.01936	.49305
1.57	.11632	.44179	2.02	.05186	.47831	2.47	.01889	.49324
1.58	.11450	.44295	2.03	.05082	.47882	2.48	.01842	.49343
1.59	.11270	.44408	2.04	.04980	.47932	2.49	.01797	.49361
1.60	.11092	.44520	2.05	.04879	.47982	2.50	.01753	.49379
1.61	.10915	.44630	2.06	.04780	.48030	2.51	.01709	.49396
1.62	.10741	.44738	2.07	.04682	.48077	2.52	.01667	.49413
1.63	.10567	.44845	2.08	.04586	.48124	2.53	.01625	.49430
1.64	.10396	.44950	2.09	.04491	.48169	2.54	.01585	.49446
1.65	.10226	.45053	2.10	.04398	.48214	2.55	.01545	.49461
1.66	.10059	.45154	2.11	.04307	.48257	2.56	.01506	.49477
1.67	.09893	.45254	2.12	.04217	.48300	2.57	.01468	.49492
1.68	.09728	.45352	2.13	.04128	.48341	2.58	.01431	.49506
1.69	.09566	.45449	2.14	.04041	.48382	2.59	.01394	.49520
1.70	.09405	.45543	2.15	.03955	.48422	2.60	.01358	.49534
1.71	.09246	.45637	2.16	.03871	.48461	2.61	.01323	.49547
1.72	.09089	.45728	2.17	.03788	.48500	2.62	.01289	.49560
1.73	.08933	.45818	2.18	.03706	.48537	2.63	.01256	.49573
1.74	.08780	.45907	2.19	.03626	.48574	2.64	.01223	.49585
1.75	.08628	.45994	2.20	.03547	.48610	2.65	.01191	.49598
1.76	.08478	.46080	2.21	.03470	.48645	2.66	.01160	.49609
1.77	.08329	.46164	2.22	.03394	.48679	2.67	.01130	.49621
1.78	.08183	.46246	2.23	.03319	.48713	2.68	.01100	.49632
1.79	.08038	.46327	2.24	.03246	.48745	2.69	.01071	.49643

t	$\phi(t)$	$\int_0^t \phi(t)dt$	t	$\phi(t)$	$\int_0^t \phi(t)dt$	t	$\phi(t)$	$\int_0^t \phi(t)dt$
2.70	.01042	.49653	3.15	.00279	.49918	3.60	.00061	.49984
2.71	.01014	.49664	3.16	.00271	.49921	3.61	.00059	.49985
2.72	.00987	.49674	3.17	.00262	.49924	3.62	.00057	.49985
2.73	.00961	.49683	3.18	.00254	.49926	3.63	.00055	.49986
2.74	.00935	.49693	3.19	.00246	.49929	3.64	.00053	.49986
2.75	.00909	.49702	3.20	.00238	.49931	3.65	.00051	.49987
2.76	.00885	.49711	3.21	.00231	.49934	3.66	.00049	.49987
2.77	.00861	.49720	3.22	.00224	.49936	3.67	.00047	.49988
2.78	.00837	.49728	3.23	.00216	.49938	3.68	.00046	.49988
2.79	.00814	.49736	3.24	.00210	.49940	3.69	.00044	.49989
2.80	.00792	.49744	3.25	.00203	.49942	3.70	.00042	.49989
2.81	.00770	.49752	3.26	.00196	.49944	3.71	.00041	.49990
2.82	.00748	.49760	3.27	.00190	.49946	3.72	.00039	.49990
2.83	.00727	.49767	3.28	.00184	.49948	3.73	.00038	.49990
2.84	.00707	.49774	3.29	.00178	.49950	3.74	.00037	.49991
2.85	.00687	.49781	3.30	.00172	.49952	3.75	.00035	.49991
2.86	.00668	.49788	3.31	.00167	.49953	3.76	.00034	.49992
2.87	.00649	.49795	3.32	.00161	.49955	3.77	.00033	.49992
2.88	.00631	.49801	3.33	.00156	.49957	3.78	.00031	.49992
2.89	.00613	.49807	3.34	.00151	.49958	3.79	.00030	.49992
2.90	.00595	.49813	3.35	.00146	.49960	3.80	.00029	.49993
2.91	.00578	.49819	3.36	.00141	.49961	3.81	.00028	.49993
2.92	.00562	.49825	3.37	.00136	.49962	3.82	.00027	.49993
2.93	.00545	.49831	3.38	.00132	.49964	3.83	.00026	.49994
2.94	.00530	.49836	3.39	.00127	.49965	3.84	.00025	.49994
2.95	.00514	.49841	3.40	.00123	.49966	3.85	.00024	.49994
2.96	.00499	.49846	3.41	.00119	.49968	3.86	.00023	.49994
2.97	.00485	.49851	3.42	.00115	.49969	3.87	.00022	.49995
2.98	.00471	.49856	3.43	.00111	.49970	3.88	.00021	.49995
2.99	.00457	.49861	3.44	.00107	.49971	3.89	.00021	.49995
3.00	.00443	.49865	3.45	.00104	.49972	3.90	.00020	.49995
3.01	.00430	.49869	3.46	.00100	.49973	3.91	.00019	.49995
3.02	.00417	.49874	3.47	.00097	.49974	3.92	.00018	.49996
3.03	.00405	.49878	3.48	.00094	.49975	3.93	.00018	.49996
3.04	.00393	.49882	3.49	.00090	.49976	3.94	.00017	.49996
3.05	.00381	.49886	3.50	.00087	.49977	3.95	.00016	.49996
3.06	.00370	.49889	3.51	.00084	.49978	3.96	.00016	.49996
3.07	.00358	.49893	3.52	.00081	.49978	3.97	.00015	.49996
3.08	.00348	.49897	3.53	.00079	.49979	3.98	.00014	.49997
3.09	.00337	.49900	3.54	.00076	.49980	3.99	.00014	.49997
3.10	.00327	.49903	3.55	.00073	.49981			
3.11	.00317	.49906	3.56	.00071	.49981			
3.12	.00307	.49910	3.57	.00068	.49982			
3.13	.00298	.49913	3.58	.00066	.49983			
3.14	.00288	.49916	3.59	.00063	.49983			

Source: Reprinted by permission from Kenney, *Mathematics of Statistics*, Part One, New York: D. Van Nostrand, pp. 225-227.

Table 4A-2
Percentiles of the χ^2 Distribution
(Column headings are percentages greater than the table values)

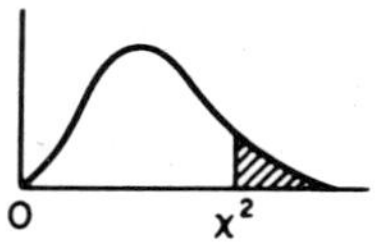

Degrees of freedom	99.5	99.0	98.0	95.0	90.0	80.0	70.0
1	0.000039	0.00016	0.00063	0.00393	0.0158	0.0642	0.148
2	0.0100	0.0201	0.0404	0.1026	0.211	0.446	0.713
3	0.0717	0.115	0.185	0.352	0.584	1.005	1.424
4	0.207	0.297	0.429	0.711	1.064	1.649	2.195
5	0.412	0.554	0.752	1.145	1.610	2.343	3.000
6	0.676	0.872	1.134	1.635	2.204	3.070	3.828
7	0.989	1.239	1.564	2.167	2.833	3.822	4.671
8	1.344	1.646	2.032	2.733	3.490	4.594	5.527
9	1.735	2.088	2.532	3.325	4.168	5.380	6.393
10	2.156	2.558	3.059	3.940	4.865	6.179	7.267
11	2.603	3.053	3.609	4.575	5.578	6.989	8.148
12	3.074	3.571	4.178	5.226	6.304	7.807	9.034
13	3.565	4.107	4.765	5.892	7.042	8.634	9.926
14	4.075	4.660	5.368	6.571	7.790	9.467	10.821
15	4.601	5.229	5.985	7.261	8.547	10.307	11.721
16	5.142	5.812	6.614	7.962	9.312	11.152	12.624
17	5.697	6.408	7.255	8.672	10.085	12.002	13.531
18	6.265	7.015	7.906	9.390	10.865	12.857	14.440
19	6.844	7.633	8.567	10.117	11.651	13.716	15.352
20	7.434	8.260	9.237	10.851	12.443	14.578	16.266
21	8.034	8.897	9.915	11.591	13.240	15.445	17.182
22	8.643	9.542	10.600	12.338	14.042	16.314	18.101
23	9.260	10.196	11.293	13.091	14.848	17.187	19.021
24	9.886	10.856	11.992	13.848	15.659	18.062	19.943
25	10.520	11.524	12.697	14.611	16.473	18.940	20.867
26	11.160	12.198	13.409	15.379	17.292	19.820	21.792
27	11.808	12.879	14.125	16.151	18.114	20.703	22.719
28	12.461	13.565	14.847	16.928	18.939	21.588	23.647
29	13.121	14.257	15.574	17.708	19.768	22.475	24.577
30	13.787	14.954	16.306	18.493	20.599	23.364	25.508
40	20.707	22.164	23.824	26.509	29.051	32.352	34.876
60	35.534	37.485	39.689	43.188	46.459	50.647	53.815
80	51.172	53.540	56.204	60.391	64.278	69.213	72.920
100	67.328	70.065	73.134	77.929	82.358	87.950	92.133
200	152.241	156.432	161.099	168.279	174.835	183.006	189.052

Source: Values in this table are reproduced from R.A. Fisher, *Statistical Methods for Research Workers* (14th ed.), with the permission of Hafner Press, New York, N.Y. (copyright © 1972 by Hafner Press.)

50.0	30.0	20.0	10.0	5.0	2.0	1.0	0.5
0.455	1.074	1.642	2.706	3.841	5.412	6.635	7.879
1.386	2.408	3.219	4.605	5.991	7.824	9.210	10.597
2.366	3.665	4.642	6.251	7.815	9.837	11.341	12.838
3.557	4.878	5.989	7.779	9.488	11.668	13.277	14.860
4.351	6.064	7.289	9.236	11.070	13.388	15.086	16.750
5.348	7.231	8.558	10.645	12.592	15.033	16.812	18.548
6.346	8.383	9.803	12.017	14.067	16.622	18.475	20.278
7.344	9.524	11.030	13.362	15.507	18.168	20.090	21.955
8.343	10.656	12.242	14.684	16.919	19.697	21.666	23.589
9.342	11.781	13.442	15.987	18.307	21.161	23.209	25.188
10.341	12.899	14.631	17.275	19.675	22.618	24.725	26.757
11.340	14.011	15.812	18.549	21.026	24.054	26.217	28.299
12.340	15.119	16.985	19.812	22.362	25.472	27.688	29.819
13.339	16.222	18.151	21.064	23.685	26.873	29.141	31.319
14.339	17.322	19.311	22.307	24.996	28.259	30.578	32.801
15.338	18.418	20.465	23.542	26.296	29.633	32.000	34.267
16.338	19.511	21.615	24.769	27.587	30.995	33.409	35.718
17.338	20.601	22.760	25.989	28.869	32.346	34.805	37.156
18.338	21.689	23.900	27.204	30.144	33.687	36.191	38.582
19.337	22.775	25.038	28.412	31.410	35.020	37.566	39.997
20.337	23.858	26.171	29.615	32.671	36.343	38.932	41.401
21.337	24.939	27.301	30.813	33.924	37.659	40.289	42.796
22.337	26.018	28.429	32.007	35.172	38.968	41.638	44.181
23.337	27.096	29.553	33.196	36.415	40.270	42.980	45.559
24.337	28.172	30.675	34.382	37.652	41.566	44.314	46.928
25.336	29.246	31.795	35.563	38.885	42.856	45.642	48.290
26.336	30.319	32.912	36.741	40.113	44.140	46.963	49.645
27.336	31.391	34.027	37.916	41.337	45.419	48.278	50.993
28.336	32.461	35.139	39.087	42.557	46.693	49.588	52.336
29.336	33.530	36.250	40.256	43.773	47.962	50.892	53.672
39.335	44.163	47.263	51.805	55.758	60.796	63.691	66.766
59.335	65.225	68.969	74.397	79.082	84.588	88.379	91.952
79.334	86.122	90.403	96.578	101.879	108.082	112.329	116.321
99.334	106.908	111.667	118.498	123.342	131.154	135.807	140.169
199.333	209.997	216.618	226.021	233.994	243.198	249.445	255.264

Table 4A-3
Values of the t Distribution

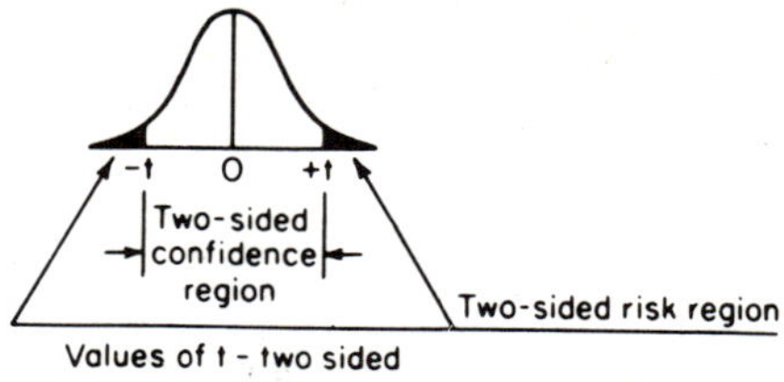

	Values of *t*—two-sided											
α risk . . .	0.90	0.80	0.70	0.60	0.50	0.40	0.30	0.20	0.10	0.05	0.02	0.01
Confidence level (prob.)	0.10	0.20	0.30	0.40	0.50	0.60	0.70	0.80	0.90	0.95	0.98	0.99
Degrees of freedom												
1	0.158	0.325	0.510	0.727	1.000	1.376	1.963	3.078	6.314	12.706	31.821	63.657
2	0.142	0.289	0.445	0.617	0.816	1.061	1.386	1.886	2.920	4.303	6.965	9.925
3	0.137	0.277	0.424	0.584	0.765	0.978	1.250	1.638	2.353	3.182	4.541	5.841
4	0.134	0.271	0.414	0.569	0.741	0.941	1.190	1.533	2.132	2.776	3.747	4.604
5	0.132	0.267	0.408	0.559	0.727	0.920	1.156	1.476	2.015	2.571	3.365	4.032
6	0.131	0.265	0.404	0.553	0.718	0.906	1.134	1.440	1.943	2.447	3.143	3.707
7	0.130	0.263	0.402	0.549	0.711	0.896	1.119	1.415	1.895	2.365	2.998	3.499
8	0.130	0.262	0.399	0.546	0.706	0.889	1.108	1.397	1.860	2.306	2.896	3.355
9	0.129	0.261	0.398	0.543	0.703	0.883	1.100	1.383	1.833	2.262	2.821	3.250
10	0.129	0.260	0.397	0.542	0.700	0.879	1.093	1.372	1.812	2.228	2.764	3.169
11	0.129	0.260	0.396	0.540	0.697	0.876	1.088	1.363	1.796	2.201	2.718	3.106
12	0.128	0.259	0.395	0.539	0.695	0.873	1.083	1.356	1.782	2.179	2.681	3.055
13	0.128	0.259	0.394	0.538	0.694	0.870	1.079	1.350	1.771	2.160	2.650	3.012
14	0.128	0.258	0.393	0.537	0.692	0.868	1.076	1.345	1.761	2.145	2.624	2.977
15	0.128	0.258	0.393	0.536	0.691	0.866	1.074	1.341	1.753	2.131	2.602	2.947
16	0.128	0.258	0.392	0.535	0.690	0.865	1.071	1.337	1.746	2.120	2.583	2.921
17	0.128	0.257	0.392	0.534	0.689	0.863	1.069	1.333	1.740	2.110	2.567	2.898
18	0.127	0.257	0.392	0.534	0.688	0.862	1.067	1.330	1.734	2.101	2.552	2.878
19	0.127	0.257	0.391	0.533	0.688	0.861	1.066	1.328	1.729	2.093	2.539	2.861
20	0.127	0.257	0.391	0.533	0.687	0.860	1.064	1.325	1.725	2.086	2.528	2.845
21	0.127	0.257	0.391	0.532	0.686	0.859	1.063	1.323	1.721	2.080	2.518	2.831
22	0.127	0.256	0.390	0.532	0.686	0.858	1.061	1.321	1.717	2.074	2.508	2.819
23	0.127	0.256	0.390	0.532	0.685	0.858	1.060	1.319	1.714	2.069	2.500	2.807
24	0.127	0.256	0.390	0.531	0.685	0.857	1.059	1.318	1.711	2.064	2.492	2.797
25	0.127	0.256	0.390	0.531	0.684	0.856	1.058	1.316	1.708	2.060	2.485	2.787
26	0.127	0.256	0.390	0.531	0.684	0.856	1.058	1.315	1.706	2.056	2.479	2.779
27	0.127	0.256	0.389	0.531	0.684	0.855	1.057	1.314	1.703	2.052	2.473	2.771
28	0.127	0.256	0.389	0.530	0.683	0.855	1.056	1.313	1.701	2.048	2.467	2.763
29	0.127	0.256	0.389	0.530	0.683	0.854	1.055	1.311	1.699	2.045	2.462	2.756
30	0.127	0.256	0.389	0.530	0.683	0.854	1.055	1.310	1.697	2.042	2.457	2.750
40	0.127	0.255	0.388	0.529	0.681	0.851	1.051	1.303	1.684	2.021	2.423	2.704
60	0.127	0.254	0.387	0.527	0.679	0.848	1.046	1.296	1.671	2.000	2.390	2.660
120	0.126	0.254	0.386	0.526	0.677	0.845	1.041	1.289	1.658	1.980	2.358	2.617
∞	0.126	0.253	0.385	0.524	0.674	0.842	1.036	1.282	1.645	1.960	2.326	2.576
Confidence level (prob.)	0.55	0.60	0.65	0.70	0.75	0.80	0.85	0.90	0.95	0.975	0.99	0.995
α risk	0.45	0.40	0.35	0.30	0.25	0.20	0.15	0.10	0.05	0.025	0.01	0.005
	Values of *t*—one-sided											

Source: Values in this table are reproduced from R.A. Fisher, *Statistical Methods for Research Workers* (14th ed.) with the permission of Hafner Press, New York, N.Y. Copyright © 1972 by Hafner Press.

5 Least-Squares Methods and Estimates

This chapter contains a general discussion of least-squares smoothing and prediction methods from an engineering point of view. The discussion starts with a description of the classical least-squares polynomial fitting technique and proceeds to a discussion of recursive smoothing and prediction equations. Recursive estimators are introduced by a description of α-β trackers where α and β are constant parameters.

Following this, the *linear* least-squares methods are discussed, using matrix notation. These methods are then extended to recursive formulations. The recursive estimators are transformed to Kalman filtering equations by the introduction of *a priori* system equations. This a priori knowledge may be obtained from physical laws governing the system or from past information covering similar types of engineering systems. The chapter also includes a numerical example describing the application of Kalman filtering equations.

Method of Least Squares

The method of least squares is used to find the closest (in a least-squares sense) polynomial approximation to a given set of data. Assume that a set of data consisting of $m + 1$ data points

$$(x_0, y_0), (x_1, y_1), (x_2, y_2), \ldots, (x_m, y_m) \tag{5.1}$$

is given. We assume a polynomial of degree n (with $n < m$) can be used to approximate these data, i.e.,

$$y = p(x) = a_n x^n + a_{n-1} x^{n-1} + \ldots + a_1 x + a_0 \tag{5.2}$$

We now proceed to evaluate the coefficients $a_n, \ldots, a_0$ by minimizing the

---

Portions of this chapter were taken from an article of the same title printed in *Proceedings of the Fifth Iranian Conference on Electrical Engineering*, Shiraz, Iran, October 27–30, 1975, pp. 1395-1423.

The author wishes to acknowledge the critical comments of B.C.C. Change, D.G. Kruska, and K.S. Jew in the preparation of this chapter. The derivation of the recursive relation of the α–β tracker given in Appendix 5A was provided by R.M. Emerick.

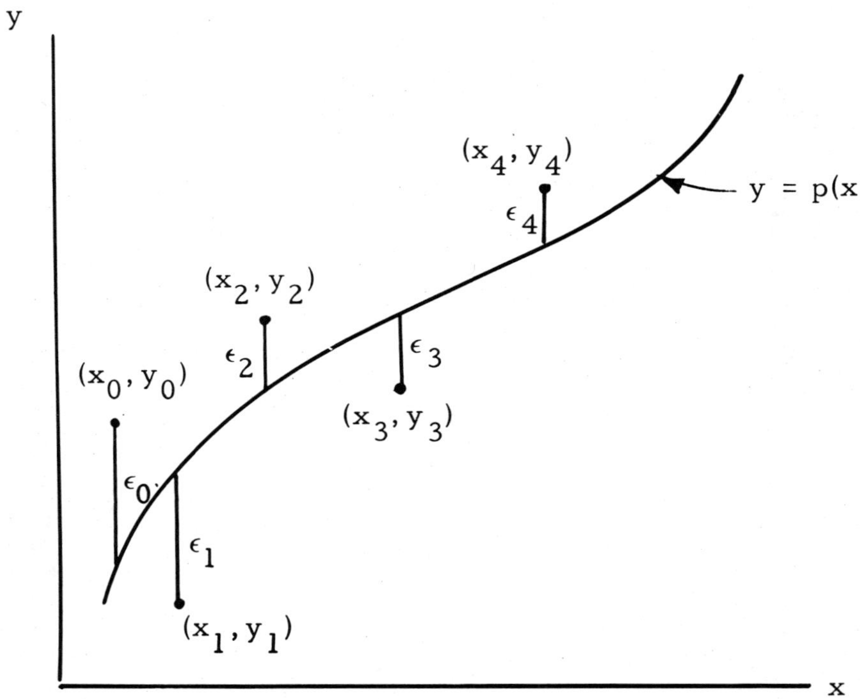

Figure 5-1. Error Between Data Points and the Fitted Curve

sum of the squares of the error in y from (5.2) as compared with actual values of $y_0, \ldots, y_m$ from (5.1). The error expression is written

$$\begin{aligned} \text{Error} = E = {} & [y_0 - (a_n x_0^n + a_{n-1} x_0^{n-1} + \ldots + a_1 x_0 + a_0)]^2 \\ & + [y_1 - (a_n x_1^n + a_{n-1} x_1^{n-1} + \ldots a_1 x_1 + a_0)]^2 \\ & + \ldots + [y_m - (a_n x_m^n + a_{n-1} x_m^{n-1} + \ldots + a_1 x_m + a_0)]^2 \end{aligned} \quad (5.3)$$

The terms of equation (5.3) represent the squares of the y-differences, ϵ_0, ϵ_1, ϵ_2, ϵ_3, . . ., of Figure 5-1.

The values of the coefficients $a_n, \ldots, a_0$ are obtained by minimizing the error E of equation (5.3). Setting the partial derivatives of E with respect to $a_n, \ldots, a_0$ equal to zero, we get

$$\begin{aligned} \frac{\partial E}{\partial a_n} = {} & 2(a_n x_0^n + a_{n-1} x_0^{n-1} + \ldots + a_0 - y_0) x_0^n \\ & + 2(a_n x_1^n + a_{n-1} x_1^{n-1} + \ldots + a_0 - y_1) x_1^n + \ldots = 0 \end{aligned} \quad (5.4)$$

and similarly for

$$\frac{\partial E}{\partial a_{n-1}}, \quad \frac{\partial E}{\partial a_{n-2}}, \quad \ldots, \quad \frac{\partial E}{\partial a_0} \tag{5.5}$$

Equations (5.4) and (5.5) can be put in the form

$$\begin{aligned} a_n[2n] + a_{n-1}[2n - 1] + \ldots + a_0[n] &= [y(n)] \\ a_n[2n - 1] + a_{n-1}[2n - 2] + \ldots + a_0[n - 1] &= [y(n - 1)] \\ &\ldots \\ a_n[n] + a_{n-1}[n - 1] + \ldots + a_0[0] &= [y(0)] \end{aligned} \tag{5.6}$$

where the notation

$$\sum_{i=0}^{m} x_i^j = [j] \quad \text{and} \quad \sum_{i=0}^{m} y_i x_i^j = [y(j)]$$

is used. Equations (5.6) can be put in matrix notation

$$\mathbf{SA} = \mathbf{T} \quad \text{and} \quad \mathbf{A} = \mathbf{S}^{-1}\mathbf{T} \tag{5.7}$$

where **A** is the matrix of the coefficients $a_n, \ldots, a_0$, **S** is the square matrix of multipliers, and **T** represents the right side of equation (5.6). The solution of matrix equation (5.7) will result in the coefficients of polynomial (5.2).

Example: Consider the four data points corresponding to equation (5.1) to be given ($m = 3$)

$$(0, 6); \quad (1, 12); \quad (2, 20); \quad (3, 35) \tag{5.8}$$

The data points are to be fitted with a second-degree polynomial ($n = 2$),

$$y = p(x) = a_2x^2 + a_1x + a_0 \tag{5.9}$$

The equations corresponding to equation (5.6) can be written

$$\begin{aligned} a_2[4] + a_1[3] + a_0[2] &= [y(2)] \\ a_2[3] + a_1[2] + a_0[1] &= [y(1)] \\ a_2[2] + a_1[1] + a_0[0] &= [y(0)] \end{aligned} \tag{5.10}$$

Equation (5.10) results in the **S**, **A**, and **T** matrices of (5.7), as

$$\begin{bmatrix} 98 & 36 & 14 \\ 36 & 14 & 6 \\ 14 & 6 & 4 \end{bmatrix} \begin{pmatrix} a_2 \\ a_1 \\ a_0 \end{pmatrix} = \begin{pmatrix} 407 \\ 157 \\ 73 \end{pmatrix} \tag{5.11}$$

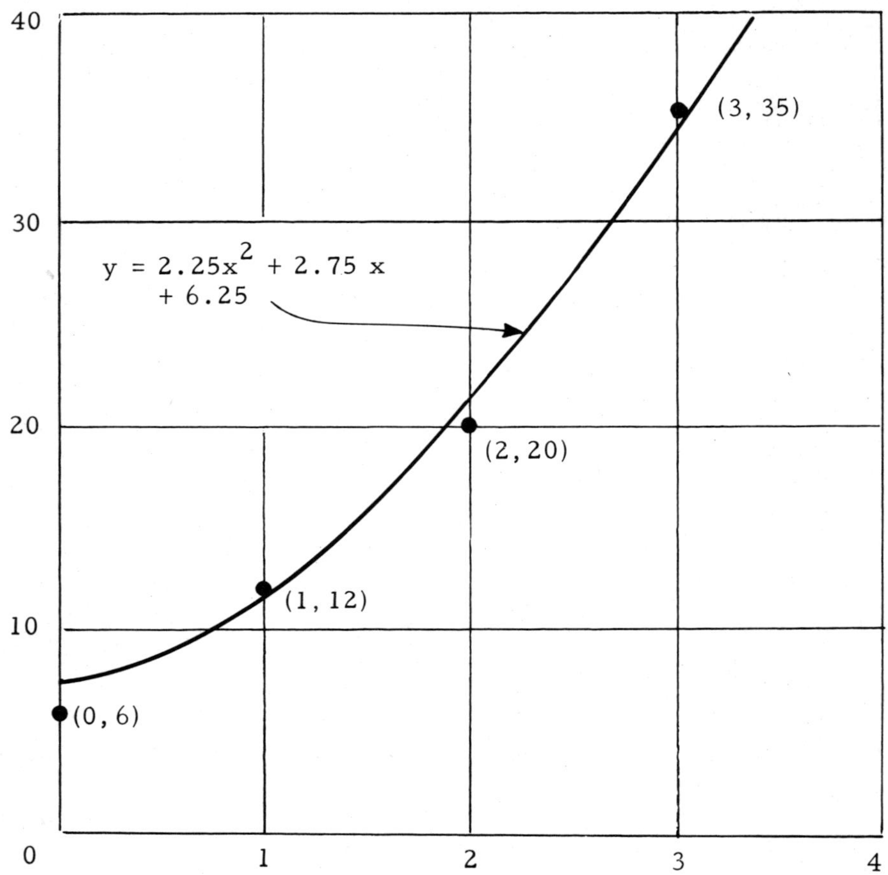

Figure 5-2. Least-Squares Fit of the Example Problem

Or, after inverting **S**, we may write

$$\begin{pmatrix} a_2 \\ a_1 \\ a_0 \end{pmatrix} = \begin{pmatrix} 0.25 & -0.75 & 0.25 \\ -0.75 & 2.45 & -1.05 \\ 0.25 & -1.05 & 0.95 \end{pmatrix} \begin{pmatrix} 407 \\ 157 \\ 73 \end{pmatrix} = \begin{pmatrix} 2.25 \\ 2.75 \\ 6.25 \end{pmatrix} \qquad (5.11a)$$

And, finally, we get the coefficients of polynomial (5.9)

$$y = 2.25x^2 + 2.75x + 6.25 \qquad (5.12)$$

Figure 5-2 gives a graph of the example in (5.12).

Discussion

The method of least squares is widely used for fitting a set of data points by a polynomial. In fact, almost all digital computer installations have library programs for the evaulation of the coefficients of the polynomial. It is seen that effectively there is no limit to the number of data points or the degree of polynomial to be fitted, the only restriction being that the number of data points should be greater than the degree of polynomial to be fitted. This requirement will result in sufficient linear simultaneous equations (5.6) for the evaluation of the coefficients of the polynomial (5.2),

A drawback of the least-squares method is the fact that with each additional data point the calculations of simultaneous linear equations (5.6) should be repeated to obtain a new set of polynomial coefficients. In other words, the calculations are *not* recursive. Examples of recursive trackers are given in the following pages.

Recursive Trackers

As an example of the linear recursive tracker, let us consider the α-β tracker[1]

$$\hat{x}_n = \alpha x_n + (1 - \alpha)x_{p_n} \tag{5.13a}$$

$$\dot{x}_n = \dot{x}_{n-1} + \frac{\beta}{T}(x_n - x_{p_n}) \tag{5.13b}$$

$$x_{p(n+1)} = \hat{x}_n + T\dot{x}_n \tag{5.13c}$$

where $\hat{x}_n$ is the smooth value of parameter x, x_n is the observation or new data point, x_{p_n} is the predicted value for time step n, dots indicate derivatives with respect to time, and T is the time step size. The parameters α and β are tracking constants. Equations (5.13) are a set of linear equations in parameter x. Note also that only the parameter x is measured and represented by observation x_n. The velocity $\dot{x}$ is calculated, not measured. Let us examine equations (5.13) in more detail.

The first equation gives the "smooth" value of the parameter x in terms of the observation made at time step n, x_n, and the predicted value of x for this time step, x_{p_n}. This, for example, means that for an α of 0.30, we put 30 percent weight on the new data (the observation x_n) and $(1 - \alpha) = 70$ percent weight on the track data or the predicted value x_{p_n}. The second equation sets the velocity of time step n equal to the velocity of time step $n - 1$ plus a correction term. This correction term $(\beta/T)(x_n - x_{p_n})$ is proportional to the difference between the observed value for time step n and the predicted value for this time step. The reason is that part of the difference between the observed and predicted values can be attributed to a velocity change. The last equation of the set merely predicts the value of x for the next time step by adding the present smooth value and the velocity/time product.

Equations (5.13) are recursive relations, as they do not involve storage of previous data; i.e., the values of parameter x for the next time step are predicted, using the values of the current time step.

Equations (5.13) can be put in the form of a *single* recursive relation (see Appendix 5A) as follows:

$$\hat{x}_n = (2 - \alpha - \beta)\hat{x}_{n-1} - (1 - \alpha)\hat{x}_{n-2} + \alpha x_n + (\beta - \alpha)x_{n-1} \quad (5.14)$$

Shifting the time index n by two time steps (replace n by $n + 2$), equation (5.14) becomes

$$\hat{x}_{n+2} - (2 - \alpha - \beta)\hat{x}_{n+1} + (1 - \alpha)\hat{x}_n = \alpha x_{n+2} + (\beta - \alpha)x_{n+1} \quad (5.15)$$

where the right side includes the observations x and the left side includes the smooth values. Note that the subscripts in (5.15) represent a shift in time; i.e.

$$\hat{x}_n = \hat{x}(t)$$

$$\hat{x}_{n+2} = \hat{x}(t + 2T) \quad (5.15a)$$

To study equations (5.15), we may use the finite difference operators which were discussed in Chapter 3 and are extensively studied in C. Jordan.[2] The finite difference operator is written

$$E^i[x(t)] = x(t + iT) \quad (5.16)$$

where i represents a shift in time t. Using (5.16) in (5.15), we get

$$[E^2 - (2 - \alpha - \beta)E + (1 - \alpha)]\hat{x}(t) = E[\alpha E + (\beta - \alpha)]x(t) \quad (5.17)$$

The complementary solution of (5.17) is obtained by finding the roots of the characteristic equation

$$q^2 - (2 - \alpha - \beta)q + (1 - \alpha) = 0$$

or

$$q = \frac{(2 - \alpha - \beta) \pm \sqrt{(2 - \alpha - \beta)^2 - 4(1 - \alpha)}}{2} \quad (5.18)$$

Using equation (5.18), we may obtain a relation between α and β parameters. Consider the case where the radical on the right of equation (5.18) vanishes, resulting in

$$\beta = \beta_{\text{critical}} = (1 - \sqrt{1 - \alpha})^2$$

$$q_1 = q_2 = \sqrt{1 - \alpha} \tag{5.19}$$

In this case, the complementary solution becomes

$$\hat{x}_{cn} = (1 - \alpha)^{n/2}\,[C_1 + C_2(n)] \tag{5.20}$$

where C_1 and C_2 are constants.

Benedict and Bordner[3] define the relationship between α and β for the "optimum" tracker as

$$\beta = \frac{\alpha^2}{2 - \alpha} \tag{5.21}$$

The optimum tracker is defined as the one resulting in the minimum value of a combination of steady-state and transient variance of the system.[4] Using (5.21) in (5.18), we get

$$\begin{aligned} q_1, q_2 &= \frac{2(1 - \alpha)}{2 - \alpha} \pm j\frac{\alpha\sqrt{1 - \alpha}}{2 - \alpha} \\ &= \alpha \pm jb \\ &= Re^{\pm j\phi} \end{aligned} \tag{5.22}$$

where

$$R = \sqrt{a^2 + b^2} = \sqrt{1 - \alpha};\ \phi = \tan^{-1}\frac{b}{a} = \tan^{-1}\frac{\alpha}{2\sqrt{1 - \alpha}} \tag{5.22a}$$

These values result in a complementary solution which is underdamped:

$$\hat{x}_{cn} = (1 - \alpha)^{n/2}[c_3 \cos(n\phi) + c_4 \sin(n\phi)] \tag{5.23}$$

From equations (5.20) and (5.23), it is seen that the complementary solutions of the *critical* and the *underdamped* systems approach zero as n assumes large values ($\alpha < 1$). These solutions correspond to the steady-state conditions. Figure 5-3 shows the critical and the underdamped behavior of the α-β tracker.

Considering the preceding derivations, we may pose an interesting question. What inputs can the α-β tracker track with no steady-state error? Denoting the input by $x(t)$, for zero steady-state error, the smooth value should also be equal to $x(t)$; i.e., $\hat{x}(t) = x(t)$. Setting these values in (5.17), we get

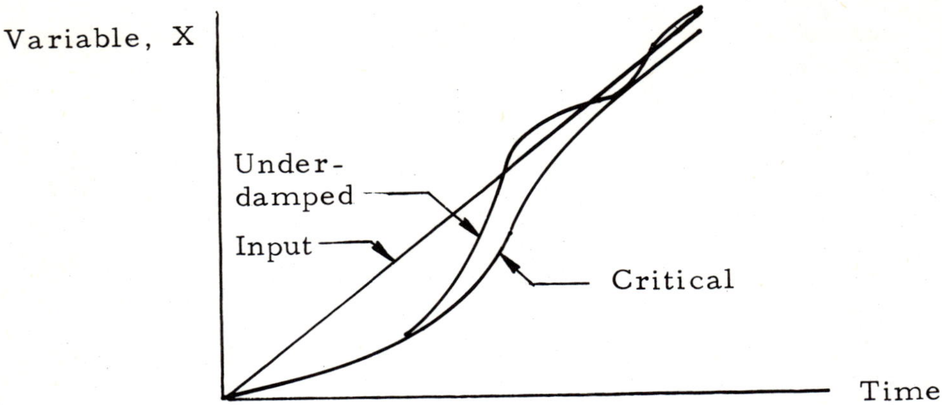

Figure 5-3. Critical and Underdamped Behavior of α-β Tracker

$$(1 - \alpha)(E^2 - 2E + 1)x = (1 - \alpha)(E - 1)^2 x = 0 \qquad (5.24)$$

The solution of this difference equation is

$$x = C_1 + C_2(nT) \qquad (5.25)$$

This implies that an α-β tracker will follow a ramp with no steady-state error. Or, in other words, if our input is

$$x = x_0 + \dot{x}_0 t \qquad (5.26)$$

the particular solution will be exactly the same as the input. Similar reasoning will show that an α-β tracker will follow the input

$$x = x_0 + \dot{x}_0 t + \ddot{x}_0 \frac{t^2}{2} \qquad (5.27)$$

which is position, velocity, acceleration, with a constant steady-state error.

The α-β-γ Tracker

An extension of the previously described α-β tracker is the α-β-γ tracker. This tracker follows a ramp plus acceleration input as given in equation (5.27), with no steady-state error. Equations of the α-β-γ tracker are as follows:

$$\hat{x}_n = \alpha x_n + (1 - \alpha) x_{p_n}$$

$$\dot{x}_n = \dot{x}_{n-1} + T\ddot{x}_{n-1} + \frac{\beta}{T}(x_n - x_{p_n})$$

$$\ddot{x}_n = \ddot{x}_{n-1} + \frac{2\gamma}{T^2}(x_n - x_{p_n})$$

$$x_{p(n+1)} = \hat{x}_n + T\dot{x}_n + \frac{1}{2}T^2\ddot{x}_n \tag{5.28}$$

The set of equations in (5.28) is similar to the previously given set of α-β tracker equations, except for the addition of acceleration terms. The set of equations (5.28) can be put in the form of a single recursion equation similar to (5.17). This recursion equation can be written in finite difference form[5] as follow:

$$[E^3 + (-3 + \alpha + \beta + \gamma)E^2 + (3 - 2\alpha - \beta + \gamma)E + (\alpha - 1)]\hat{x}(t)$$
$$= E\,[\alpha E^2 + (-2\alpha + \beta + \gamma)E + (\alpha - \beta + \gamma)]x(t) \tag{5.29}$$

The complementary solution can be obtained from equation (5.29). In fact, the values of γ and β in terms of α for the optimum tracker are expressed in Simpson.[6] Using these relations, one can set the characteristic equation of (5.29) in terms of parameter α only and obtain the complementary solution.

Equations (5.28) can be put in matrix notation:

$$\underset{(3\times 1)}{\hat{\mathbf{X}}_{n+1}} = \underset{(3\times 3)(3\times 1)}{\phi\hat{\mathbf{X}}_n} + \underset{(3\times 1)}{\mathbf{K}}\ \underset{(1\times 1)}{(x_n - \hat{x}_n)} \tag{5.30}$$

where

$$\hat{\mathbf{X}}_{n+1} = \begin{pmatrix} \hat{x}_{n+1} \\ \dot{x}_{n+1} \\ \ddot{x}_{n+1} \end{pmatrix}; \quad \hat{\mathbf{X}}_n = \begin{pmatrix} \hat{x}n \\ \dot{x}_n \\ \ddot{x}_n \end{pmatrix}$$

$$\phi = \begin{pmatrix} 1 & T & T^2/2 \\ 0 & 1 & T \\ 0 & 0 & 1 \end{pmatrix}; \quad \mathbf{K} = \begin{pmatrix} \alpha \\ \beta/T \\ 2\gamma/T^2 \end{pmatrix}$$

and $(x_n - \hat{x}_n)$ = one-by-one matrix. Equation (5.30) may be considered as the predicted values for $(n + 1)$ in terms of current (n) values. In equation (5.30), $\hat{\mathbf{X}}_{n+1}$ is the predicted vector for the next $(n + 1)$ time step, ϕ is the transition matrix, **K** is the matrix of the coefficients, and x_n is the value of observation at time step n. Note that the right side of equation (5.30) involves terms at the nth time step, while the left side involves terms of the $n + 1$ time step. The matrix representation of the α-β-γ tracker will be used in future discussions.

Linear Least-Squares Estimates

In linear least-squares methods, it is assumed that a *linear* relationship exists between the independent variable **X** and the dependent variable **Y**. This relationship can be expressed in matrix notation as follows.

$$\underset{(m\times 1)}{\mathbf{Y}} = \underset{(m\times n)}{\mathbf{W}} \; \underset{(n\times 1)}{\mathbf{X}} \tag{5.31}$$

where

$$\mathbf{Y} = \begin{pmatrix} y_1 \\ y_2 \\ \cdots \\ y_m \end{pmatrix}; \quad \mathbf{W} = \begin{pmatrix} w_{11} & w_{12} & \ldots\, w_{1n} \\ w_{21} & w_{22} & \ldots\, w_{2n} \\ \cdots & & \\ w_{m1} & w_{m2} & w_{mn} \end{pmatrix}; \quad \mathbf{X} = \begin{pmatrix} x_1 \\ x_2 \\ \cdots \\ x_n \end{pmatrix}$$

Note that (5.31) represents a linear relation, as the elements of the matrix are not a function of the independent variable x. Consider the case where a set of measurements $y_1, y_2, y_3, \ldots, y_n$ are obtained and it is desired to calculate $\hat{\mathbf{X}}$, minimizing the least-squares cost function.

$$\mathbf{J} = \frac{1}{2}(\mathbf{Y} - \mathbf{W}\hat{\mathbf{X}})' \, (\mathbf{Y} - \mathbf{W}\hat{\mathbf{X}}) \tag{5.32}$$

where primes denote transposed matrices. The cost function of equation (5.32) represents the least-squares sum of errors as shown by the following discussion. Note that the matrix (**Y** − **WX**) basically represents an error matrix. Its elements will be as follows:

$$\begin{pmatrix} y_1 - (w_{11}x_1 + w_{12}x_2 + \ldots + w_{1n}x_n) \\ y_2 - (w_{21}x_1 + w_{22}x_2 + \ldots + w_{2n}x_n) \\ \cdots \\ y_n - (w_{n1}x_1 + w_{n2}x_2 + \ldots + w_{nn}x_n) \end{pmatrix} = \begin{pmatrix} \Delta\epsilon_1 \\ \Delta\epsilon_2 \\ \cdots \\ \Delta\epsilon_n \end{pmatrix} \tag{5.33}$$

where $\Delta\epsilon$ terms designate the error terms. Using the relation of (5.33), equation (5.32) becomes

$$\mathbf{J} = \frac{1}{2}(\Delta\epsilon_1, \Delta\epsilon_2, \ldots, \Delta\epsilon_n)\begin{pmatrix} \Delta\epsilon_1 \\ \Delta\epsilon_2 \\ \vdots \\ \Delta\epsilon_n \end{pmatrix} \quad (5.34)$$

or we get

$$\mathbf{J} = \frac{1}{2}(\Delta\epsilon_1^2 + \Delta\epsilon_2^2 + \Delta\epsilon_3^2 + \ldots + \Delta\epsilon_n^2)$$

representing the sum square of error terms similar to equation (5.3).

Equation (5.32) can be written as

$$\mathbf{J} = \frac{1}{2}(\mathbf{Y}' - \hat{\mathbf{X}}'\mathbf{W}')(\mathbf{Y} - \mathbf{W}\hat{\mathbf{X}}) \quad (5.35)$$

$$\mathbf{J} = \frac{1}{2}[\mathbf{Y}\mathbf{Y}' - \mathbf{Y}'\mathbf{W}\hat{\mathbf{X}} - \hat{\mathbf{X}}'\mathbf{W}'\mathbf{Y} + \hat{\mathbf{X}}'\mathbf{W}'\mathbf{W}\hat{\mathbf{X}}]$$

Since matrix products $\mathbf{Y}'\mathbf{W}\hat{\mathbf{X}} = \hat{\mathbf{X}}'\mathbf{W}'\mathbf{Y}$, we get

$$\mathbf{J} = \frac{1}{2}(\mathbf{Y}'\mathbf{Y}) - (\mathbf{Y}'\mathbf{W}\hat{\mathbf{X}}) + \frac{1}{2}\hat{\mathbf{X}}'\mathbf{W}'\mathbf{W}\hat{\mathbf{X}}$$

(The products $\mathbf{Y}'\mathbf{W}\hat{\mathbf{X}}$, and $\hat{\mathbf{X}}'\mathbf{W}'\mathbf{Y}$ will be equal to a one-element matrix representing the $\mathbf{Y}'\mathbf{W}\hat{\mathbf{X}}$ and its transpose $\hat{\mathbf{X}}'\mathbf{W}'\mathbf{Y}$.)

Obtaining the derivative with respect to $\hat{\mathbf{X}}$ and setting the same equal to zero, we get

$$\frac{\partial \mathbf{J}}{\partial \hat{\mathbf{X}}} = -\mathbf{W}'\mathbf{Y} + \mathbf{W}'\mathbf{W}\hat{\mathbf{X}} = 0$$

From this we have

$$\hat{\mathbf{X}} = [\mathbf{W}'\mathbf{W}]^{-1}\,\mathbf{W}'\mathbf{Y} \quad (5.36)$$

Example

As an example of the preceding equations, consider the polynomial fit of equation (5.2) to be put in the form of equation (5.31).

$$\mathbf{Y} = \mathbf{W}\mathbf{X}$$

where

$$\mathbf{Y} = \begin{pmatrix} y_o \\ y_1 \\ \vdots \\ y_m \end{pmatrix}; \quad \mathbf{W} = \begin{pmatrix} x_o^n & x_o^{n-1} & \dots & x_o & 1 \\ x_1^n & x_1^{n-1} & \dots & x_1 & 1 \\ \dots & & & & \\ x_m^n & x_m^{n-1} & \dots & x_m & 1 \end{pmatrix}$$

$$\mathbf{X} = \begin{pmatrix} a_n \\ a_{n-1} \\ \vdots \\ a_o \end{pmatrix}$$

The estimate $\hat{\mathbf{X}}$ of (5.36) corresponds to the coefficient matrix **A** of (5.7) and, additionally, $[\mathbf{W}'\mathbf{W}] = \mathbf{S}$ and $\mathbf{W}'\mathbf{Y} = \mathbf{T}$. From this point of view ys are the measurements of dependent variable and as are the independent variable to be estimated using least-squares errors. If variables x of the polynomial (5.2) were considered as the independent variables (instead of as) then the polynomial fit will express a nonlinear relationship.

Gauss-Markov Theorem

Now, instead of equation (5.31), consider the case where the relationship between dependent and independent variables **Y** and **X** also includes uncorrelated noise with zero mean and variance **Q**, as given below.

$$\mathbf{Y} = \mathbf{W}\mathbf{X} + \epsilon \tag{5.37}$$

with $\mathbf{E}(\epsilon) = 0$ and $\mathbf{E}(\epsilon\epsilon') = \mathbf{Q}$, where $\mathbf{E}(\epsilon)$ is the *expected* value and $\mathbf{E}(\epsilon\epsilon')$ is the *covariance*. In this case, we are required to find $\hat{\mathbf{X}}$ subject to the minimization of the cost function.

$$\mathbf{J} = \frac{1}{2}(\mathbf{Y} - \mathbf{W}\hat{\mathbf{X}})'\mathbf{Q}^{-1}(\mathbf{Y} - \mathbf{W}\hat{\mathbf{X}}) \tag{5.38}$$

Note the introduction of weighting function **Q** in the cost function as compared to the previous cost function of equation (5.32). Observe that the weighting function **Q** represents the noise covariance of the measurement **Y** of (5.37).

Following the same type of derivation as discussed for the minimization of the previously given cost function (5.32) for the weighted cost function (5.38), we get

$$\hat{\mathbf{X}} = [\mathbf{W}'\mathbf{Q}^{-1}\mathbf{W}]^{-1}\,\mathbf{W}'\mathbf{Q}^{-1}\mathbf{Y} \tag{5.39}$$

Denoting

$$\mathbf{P} = [\mathbf{W}'\mathbf{Q}^{-1}\mathbf{W}]^{-1} \tag{5.40}$$

we get

$$\hat{\mathbf{X}} = \mathbf{P}\mathbf{W}'\mathbf{Q}^{-1}\mathbf{Y} \tag{5.41}$$

Note that $\hat{\mathbf{X}}$ in the above equations represents the best estimate of variable $\mathbf{X}$ under the linear matrix relation (5.37).

Recursive Estimators

It is always desirable to include new data in existing equations without "starting all over" with the calculations. This was described by going from least-squares polynomial fits to the equations of α-β trackers. It was seen that the α-β coefficients were taken as *constants* for computation purposes. With the discussion of linear estimates given above, we may now proceed to derive equations of recursive linear estimators with variable coefficients (Kalman filters).

Suppose that an estimator $\hat{\mathbf{X}}$ of the variable $\mathbf{X}$ is obtained, with the resulting error variance of

$$\mathbf{E}[(\mathbf{X} - \hat{\mathbf{X}})(\mathbf{X} - \hat{\mathbf{X}})'] = \mathbf{R} \tag{5.42}$$

Given additional measurements of the form

$$\mathbf{Y} = \mathbf{W}\mathbf{X} + \epsilon \tag{5.43}$$

with $\mathbf{E}(\epsilon) = 0$, $\mathbf{E}(\epsilon\epsilon') = \mathbf{Q}$, we want to update the estimate $\hat{\mathbf{X}}$ to its new value $\hat{\hat{\mathbf{X}}}$. Note that ϵ is uncorrelated with both $\mathbf{X}$ and the past measurements.

The best estimate of $\mathbf{Y}$ based on past measurements is $\hat{\mathbf{Y}} = \mathbf{W}\hat{\mathbf{X}}$ and we may write the difference between the new measurement $\mathbf{Y}$ and $\hat{\mathbf{Y}}$ as

$$\Delta\mathbf{Y} = \mathbf{Y} - \mathbf{W}\hat{\mathbf{X}} \tag{5.44}$$

Intuitively, we may write the new estimate $\hat{\hat{\mathbf{X}}}$ of the variable $\mathbf{X}$ as a linear function of the present estimate $\hat{\mathbf{X}}$ and the correction term $\Delta\mathbf{Y}$

$$\hat{\hat{\mathbf{X}}} = \hat{\mathbf{X}} + \mathbf{D}\Delta\mathbf{Y} \tag{5.45}$$

where $\mathbf{D}$ represents the "fraction" of the correction $\Delta\mathbf{Y}$ to be incorporated in $\hat{\hat{\mathbf{X}}}$ due to measurement (5.43).

To obtain a relation for $\mathbf{D}$, consider the vector diagram of Figure 5-4. The figure gives a vector representing variable $\mathbf{X}$ and the correction term

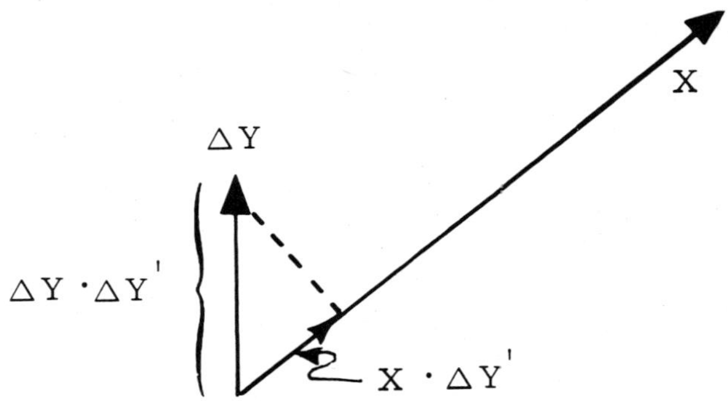

Figure 5-4. Geometric Interpretation of Linear Estimate

$\mathbf{\Delta Y}$. Note that we intend to incorporate only the portion of the correction $\mathbf{\Delta Y}$ which is in the direction of the vector $\mathbf{X}$. Translating this direction correction into the previously given notation, we may write

$$\mathbf{D} = \mathbf{E}(\mathbf{X}\,\mathbf{\Delta Y}')\,[\mathbf{E}(\mathbf{\Delta Y}\,\mathbf{\Delta Y}')]^{-1}$$

where the above represents the "fraction" of the correction in the direction of $\mathbf{X}$ as compared to the total "length" of the correction $\mathbf{\Delta Y}$. Substituting in (5.45), we get

$$\hat{\hat{\mathbf{X}}} = \hat{\mathbf{X}} + \mathbf{E}(\mathbf{X}\,\mathbf{\Delta Y}')[\mathbf{E}(\mathbf{\Delta Y}\,\mathbf{\Delta Y}')]^{-1}\,\mathbf{\Delta Y} \tag{5.46}$$

The terms of equation (5.46) are evaluated as follows:

$$\begin{aligned}\mathbf{E}(\mathbf{X}\,\mathbf{\Delta Y}') &= \mathbf{E}[\mathbf{X}(\mathbf{Y} - \mathbf{W}\hat{\mathbf{X}})'] \\ &= \mathbf{E}[\mathbf{X}(\mathbf{X}'\mathbf{W}' + \epsilon' - \hat{\mathbf{X}}'W')] \\ &= \mathbf{E}[\mathbf{X}(\mathbf{X} - \hat{\mathbf{X}})'\mathbf{W}'] \qquad (5.46a)\end{aligned}$$

where the fact of independence of $\mathbf{X}$ and ϵ in the form of $\mathbf{E}(\mathbf{X}\epsilon') = 0$ has been used. Following, we get

$$\begin{aligned}\mathbf{E}(\mathbf{X}\,\mathbf{\Delta Y}') &= \mathbf{E}[(\mathbf{X} - \hat{\mathbf{X}} + \hat{\mathbf{X}})(\mathbf{X} - \hat{\mathbf{X}})'\mathbf{W}'] \\ &= \mathbf{E}[(\mathbf{X} - \hat{\mathbf{X}})(\mathbf{X} - \hat{\mathbf{X}})'\mathbf{W}'] + \mathbf{E}[\hat{\mathbf{X}}(\mathbf{X} - \hat{\mathbf{X}})'\mathbf{W}'] \qquad (5.46b)\end{aligned}$$

Considering the fact that the difference $(\mathbf{X} - \hat{\mathbf{X}})$ is uncorrelated with $\hat{\mathbf{X}}$, the last term on the right vanishes and using (5.42) we get

$$\mathbf{E}(\mathbf{X}\mathbf{\Delta Y}') = \mathbf{R}\mathbf{W}' \tag{5.47}$$

We also may compute the term

$$\mathbf{E}(\Delta\mathbf{Y}\,\Delta\mathbf{Y}') = \mathbf{E}[(\mathbf{Y} - \mathbf{W}\hat{\mathbf{X}})(\mathbf{Y} - \mathbf{W}\hat{\mathbf{X}})']$$
$$= \mathbf{E}\{[\mathbf{W}(\mathbf{X} - \hat{\mathbf{X}}) + \epsilon]\,[\mathbf{W}(\mathbf{X} - \hat{\mathbf{X}}) + \epsilon]'\}$$
$$= \mathbf{WRW}' + \mathbf{Q}$$

Using these relations in (5.46), we get

$$\hat{\hat{\mathbf{X}}} = \hat{\mathbf{X}} + \mathbf{RW}'[\mathbf{WRW}' + \mathbf{Q}]^{-1}(\mathbf{Y} - \mathbf{W}\hat{\mathbf{X}}) \tag{5.48}$$

It can be shown that this estimate will have an error covariance matrix

$$\mathbf{E}[(\mathbf{X} - \hat{\hat{\mathbf{X}}})(\mathbf{X} - \hat{\hat{\mathbf{X}}})'] = \mathbf{R} - \mathbf{RW}'(\mathbf{WRW}' + \mathbf{Q})^{-1}\mathbf{WR} \tag{5.49}$$

The error matrix of (5.49) is related to the previously given error matrix **R** of (5.42).

Recursive Estimation and a Priori Knowledge

In the previous analysis, we relied on the data without the use of any "inside information." This inside information or a priori knowledge may come from physical laws and/or from past performance information on similar types of systems. This knowledge can be put in equation form and used with the estimation theory equations. With theory derived thus far, we are in a position to incorporate a priori knowledge in our equations. This will enable us to obtain better estimates for future predictions as well as to handle the problem of missing observations.

Definitions

Consider a system, the behavior of which can be represented by the matrix equation

$$\mathbf{X}_{n+1} = \phi_n \mathbf{X}_n + \mathbf{U}_n \tag{5.50}$$

where the index n is the time index, ϕ_n is the *transition matrix*, and $\mathbf{U}_n$ is a random vector with a mean of zero and a covariance of $\mathbf{E}[\mathbf{U}_n \cdot \mathbf{U}_n'] = \mathbf{Q}_n$. Note that the behavior at each future time step is predicted using the value of **X** at the present time step.

Assume an initial vector $\mathbf{X}_0$ together with an initial estimate $\hat{\mathbf{X}}_0$ having the covariance

$$\mathbf{E}[(\hat{\mathbf{X}}_0 - \mathbf{X})(\hat{\mathbf{X}}_0 - \mathbf{X})'] = \mathbf{P}_0 \tag{5.51}$$

Assume measurements of the form

$$\mathbf{Y}_n = \mathbf{M}_n\mathbf{X}_n + \mathbf{W}_n \tag{5.52}$$

where $\mathbf{M}$ is the measurement matrix and $\mathbf{W}_n$ is the measurement noise, are obtained at every time step n. Denote the covariance of $\mathbf{W}_n$ by $\mathbf{E}[\mathbf{W}_n\mathbf{W}_n']$ $= \mathbf{R}_n$ with $\mathbf{W}_n$ having a mean of zero.

In addition, we assume that random vectors $\mathbf{X}$, $\mathbf{U}$, and $\mathbf{W}$ are all uncorrelated.

Kalman's Theorem

Theorem: The optimal estimate of random vector $\mathbf{X}$ may be generated recursively according to the equation

$$\hat{\mathbf{X}}_{n+1} = \underbrace{\phi\hat{\mathbf{X}} + \phi\mathbf{P}\mathbf{M}'[\mathbf{M}\mathbf{P}\mathbf{M}' + \mathbf{R}]^{-1}(\mathbf{Y} - \mathbf{M}\hat{\mathbf{X}})}_{\text{All evaluated at } n} \tag{5.53}$$

where the matrix $\mathbf{P}$ is the covariance matrix, generated recursively by the equation

$$\mathbf{P}_{n+1} = \underbrace{\phi\mathbf{P}\{\mathbf{I} - \mathbf{M}'[\mathbf{M}\mathbf{P}\mathbf{M}' + \mathbf{R}]^{-1}\mathbf{M}\mathbf{P}\}\phi' + \mathbf{Q}}_{\text{All evaluated at } n} \tag{5.54}$$

The initial conditions are the initial estimates $\hat{\mathbf{X}}_0$ and $\mathbf{P}_0$.

Proof

Assume that the values of $y_0, y_1, \ldots, y_{n-1}$ are measured and that the estimate $\hat{\mathbf{X}}_n$ and the corresponding covariance matrix $\mathbf{P}_n = \mathbf{E}[(\mathbf{X}_n - \hat{\mathbf{X}}_n)(\mathbf{X}_n - \hat{\mathbf{X}}_n)']$ have been computed. At time n, we receive a new set of measurements

$$\mathbf{Y}_n = \mathbf{M}_n\mathbf{X}_n + \mathbf{W}_n \tag{5.55}$$

giving us additional information. This is the exact problem which was solved by equation (5.48). Using this equation, we get

$$\hat{\hat{\mathbf{X}}}_n = \underbrace{\hat{\mathbf{X}} + \mathbf{P}\mathbf{M}'[\mathbf{M}\mathbf{P}\mathbf{M}' + \mathbf{R}]^{-1}(\mathbf{Y} - \mathbf{M}\hat{\mathbf{X}})}_{\text{All evaluated at } n} \tag{5.56}$$

with the covariance from (5.49) of

$$\hat{\mathbf{P}}_n = \underbrace{\mathbf{P} - \mathbf{P}\mathbf{M}'[\mathbf{M}\mathbf{P}\mathbf{M}' + \mathbf{R}]^{-1}\mathbf{M}\mathbf{P}}_{\text{All evaluated at } n} \tag{5.57}$$

Using equation (5.50), we obtain the estimate of variable **X** for the $(n + 1)$th time step.

$$\hat{\mathbf{X}}_{n+1} = \phi_n \hat{\hat{\mathbf{X}}}_n \tag{5.58}$$

Substitution of $\hat{\hat{\mathbf{X}}}_n$ from (5.56) in (5.58) results in estimation equation (5.53). The error covariance of estimation (5.58) is given below

$$\mathbf{P}_{n+1} = \phi_n \hat{\mathbf{P}}_n \phi_n' + \mathbf{Q}_n \tag{5.59}$$

Substitution of $\hat{\mathbf{P}}_n$ from (5.57) in (5.59) will result in recursion equation (5.54).

Comparison of the α-β-γ tracking equation (5.30) and the Kalman filter equation (5.53) points out the similarities of these tracking systems. In equation (5.30), the components of matrix **K** remain constant, but in equation (5.53) these components are evaluated at each time step n. In going from (5.30) to (5.53), the matrix **K** transforms into

$$\mathbf{K} \rightarrow \phi \mathbf{P}\mathbf{M}'[\mathbf{M}\mathbf{P}\mathbf{M}' + \mathbf{R}]^{-1} \tag{5.60}$$

The right side of this equation incorporates the effects of error covariance matrix **P**, the transition matrix ϕ, and the measurement error matrix **R**.

Numerical Example of Kalman Filtering

Consider the problem where the distance between a given radar and the target is measured by the radar at discrete intervals of time. Assume the radar-target distance as a function of time to be given by

$$\begin{aligned} R &= R_0 + \dot{R}t \\ &= 60000 - 250t \end{aligned} \tag{5.61}$$

where R is the radar-target range in meters, R_0 is the initial range at $t = 0$, and $\dot{R}$ is the closing rate in meters/second. The closing rate $\dot{R}$ is assumed to be constant. We further assume that the radar-target range as measured by the radar will have an error with a Gaussian distribution of zero mean and a standard deviation of 200 meters. This error may be simulated in a computer program[7]

$$\epsilon_n = \left(\sum_{i=1}^{12} y_i - 6\right)(200) \tag{5.62}$$

In (5.62), y_i is a random number between 0 and 1 generated from a uniform distribution and n is the discrete time index.

If the range data are obtained at time increments of T, this can be simulated, using (5.61) and (5.62), as

$$R_n = R_0 + \dot{R}(n)(T) + \epsilon_n \tag{5.63}$$

Thus at every time step n we can obtain a value for radar-target range R_n.

Let us assume that range data are obtained every 2 seconds from the radar, i.e., $T = 2$. With this value, we can formulate the problem numerically. The matrices corresponding to (5.50) from Newton's laws of motion can be written as

$$\mathbf{X}_{n+1} = \phi_n \mathbf{X}_n + \mathbf{U}_n \tag{5.64}$$

with

$$\mathbf{X} = \begin{pmatrix} R \\ \dot{R} \\ \ddot{R} \end{pmatrix}; \quad \phi = \begin{pmatrix} 1 & T & T^2/2 \\ 0 & 1 & T \\ 0 & 0 & 1 \end{pmatrix} = \begin{pmatrix} 1 & 2 & 2 \\ 0 & 1 & 2 \\ 0 & 0 & 1 \end{pmatrix} \tag{5.65}$$

and the covariance of the $\mathbf{U}_n$ matrix

$$\mathbf{E}(\mathbf{U}_n \mathbf{U}_n') = \mathbf{Q}_n = \begin{pmatrix} 0 & 0 & 0 \\ 0 & 0 & 0 \\ 0 & 0 & 25 \end{pmatrix} \tag{5.66}$$

In defining the $\mathbf{Q}_n$ matrix, it is assumed that the acceleration term $\ddot{R}$ may vary with a variance $\sigma = 5$ meter/second^2 ($\sigma^2 = 25$), while R and $\dot{R}$ terms are expected, in general, to behave according to relation (5.63).

Since only the value of distance R is measured, matrix $\mathbf{M}_n$ of (5.52) can be written

$$\mathbf{M}_n = (1 \quad 0 \quad 0) \tag{5.67}$$

From the assumed value of Gaussian noise on measurements of range, the covariance of $\mathbf{W}_n$ of (5.52) is written

$$\mathbf{E}(\mathbf{W}_n \mathbf{W}_n') = \mathbf{R}_n = (200)^2 = 40{,}000 \tag{5.68}$$

representing the variance σ^2 of range measurements. The value of $\mathbf{R}_n$ in (5.68) should not be confused with range R_n of (5.63). The initial vectors $\mathbf{X}_0$ and $\mathbf{Y}_0$ are assumed to be

$$\mathbf{X}_0 = \begin{pmatrix} 60200 \\ -300 \\ 0 \end{pmatrix}; \quad \mathbf{Y}_0 = 60200 \tag{5.69}$$

where the value of -300 is an estimate of closing rate $\dot{R}$. The measured value of $\mathbf{Y}_0$ is used in the $\mathbf{X}_0$ for the initial estimate of radar-target range R. The matrix of covariance $\mathbf{P}_0$ of (5.51) is estimated as follows:

$$\mathbf{P}_0 = \begin{pmatrix} 40{,}000 & 0 & 0 \\ 0 & 2500 & 0 \\ 0 & 0 & 25 \end{pmatrix} \tag{5.70}$$

where 40,000, 2500, and 25 are the variances in the initial estimates of vector $\mathbf{X}_0$.

In this example, error matrices $\mathbf{Q}_n$ and $\mathbf{R}_n$ are assumed *not* to vary with time. In general, these matrices may be time varying, as the values of variances used in them may depend on time.

Having the above values at time zero, we may proceed to compute matrix expressions (5.53) and (5.54) of Kalman filtering. These values for equation (5.53) are given below.

$$(\phi\mathbf{P}\mathbf{M}')_0 = \begin{pmatrix} 40{,}000 \\ 0 \\ 0 \end{pmatrix} \tag{5.71}$$

$$[\mathbf{M}\mathbf{P}\mathbf{M}' + \mathbf{R}]_0 = 80{,}0000\,; \quad [\mathbf{M}\mathbf{P}\mathbf{M}' + \mathbf{R}]_0^{-1} = 1.25 \times 10^{-5} \tag{5.72}$$

$$(\phi\mathbf{P}\mathbf{M}')_0[\mathbf{M}\mathbf{P}\mathbf{M}' + \mathbf{R}]_0^{-1} = \begin{pmatrix} 0.5 \\ 0 \\ 0 \end{pmatrix} \tag{5.73}$$

Note that the above matrix is equivalent to the $\mathbf{K}$ matrix of the α-β-γ tracker of equation (5.30). We also have

$$(\mathbf{Y} - \mathbf{M}\mathbf{X})_0 = 0\,; \quad (\phi\mathbf{X})_0 = \begin{pmatrix} 59600 \\ -300 \\ 0 \end{pmatrix} \tag{5.74}$$

and, finally, matrix

$$\hat{\mathbf{X}}_1 = \begin{pmatrix} 59600 \\ -300 \\ 0 \end{pmatrix} \tag{5.75}$$

For the computation of $\mathbf{P}_1$ having $\mathbf{P}_0$, we proceed with equation (5.54):

$$(\mathbf{MPM}' + \mathbf{R}) = 80000 \tag{5.76}$$

$$[\mathbf{MPM}' + \mathbf{R}]^{-1} = 1.25 \times 10^{-5} \tag{5.77}$$

$$\mathbf{M}'[\mathbf{MPM}' + \mathbf{R}]^{-1}\mathbf{MP} = \begin{pmatrix} 0.5 & 0 & 0 \\ 0 & 0 & 0 \\ 0 & 0 & 0 \end{pmatrix} \tag{5.78}$$

$$\phi\mathbf{P}\{\ \}\phi' = \begin{pmatrix} 30100 & 5100 & 50 \\ 5100 & 2600 & 50 \\ 50 & 50 & 25 \end{pmatrix} \tag{5.79}$$

where { } refer to those of equation (5.54). Adding matrix **Q** of (5.66) to the matrix of (5.79), we get

$$\mathbf{P}_1 = \begin{pmatrix} 30100 & 5100 & 50 \\ 5100 & 2600 & 50 \\ 50 & 50 & 50 \end{pmatrix} \tag{5.80}$$

Now we have the estimates of $\mathbf{X}_1$ and $\mathbf{P}_1$ for time step 1. Assuming we obtain the data point

$$\mathbf{Y}_1 = 59305.7 \tag{5.81}$$

we can proceed to calculate $\mathbf{X}_2$ and $\mathbf{P}_2$ from (5.53) and (5.54), respectively. We may also calculate the **K** matrix of α-β-γ coefficients.

$$\mathbf{K}_1 = \phi\mathbf{PM}'[\mathbf{MPM}' + \mathbf{R}]^{-1} = \begin{pmatrix} 0.5763 \\ 7.42 \times 10^{-2} \\ 7.13 \times 10^{-4} \end{pmatrix}$$

$$\hat{\mathbf{X}}_2 = \begin{pmatrix} 58830.4 \\ -321.834 \\ -0.209 \end{pmatrix} ; \mathbf{P}_2 = \begin{pmatrix} 38416 & 7903 & 221 \\ 7903 & 2614 & 146 \\ 221 & 146 & 74 \end{pmatrix} \tag{5.82}$$

Similarly, with $\mathbf{Y}_2 = 59352.7$, we get

$$\mathbf{K}_2 = \begin{pmatrix} 0.6971 \\ 0.1064 \\ 2.8206 \times 10^{-3} \end{pmatrix} \tag{5.83}$$

$$\hat{\mathbf{X}}_3 = \begin{pmatrix} 58550.4 \\ -266.663 \\ 1.2634 \end{pmatrix} \tag{5.84}$$

$$\mathbf{P}_3 = \begin{pmatrix} 44733 & 8934 & 509 \\ 8934 & 2611 & 273 \\ 509 & 273 & 99 \end{pmatrix} \tag{5.85}$$

Figure 5-5 gives a graph of filtered values of R, $\dot{R}$, and $\ddot{R}$ as a function of time. The values of observations $\mathbf{Y}$ are also shown in the figure. It is noted that the values of velocity, because of the error in the initial $\dot{R}$ (300 meters/second as compared to 250 meters/second) are not converging to the actual values as rapidly as the values of distance.

Figure 5-6 is a plot of $\mathbf{K}$ vector, giving the values of the equivalent α-β-γ tracker parameters. It may be recalled that, in the α-β-γ trackers, these parameters are usually assumed to be constant. In cases where it is easier to program α-β-γ tracker equations for computers, matrix $\mathbf{K}$ can be evaluated and the values used in the α-β-γ tracker equations (5.28).

Problems

1. Consider the data points

$$x = 0,\ 2,\ 4,\ 9,\ 12$$

$$y = 0,\ 8,\ 14,\ 32,\ 60$$

Calculate the coefficients of least-squares fit to the polynomial

$$y = Ax + B$$

2. Calculate the coefficients of the least-squares fit to the polynomial

$$y = Ax^2 + Bx + C$$

using the data of problem 1.

3. Write the set of simultaneous equations for the coefficient of the polynomial

$$y = Ax^2 + Bx + C$$

to least-squares fit the data

$$x = 1,\ 3,\ 5,\ 7,\ 9$$

$$y = 0,\ 8,\ 12,\ 14,\ 22$$

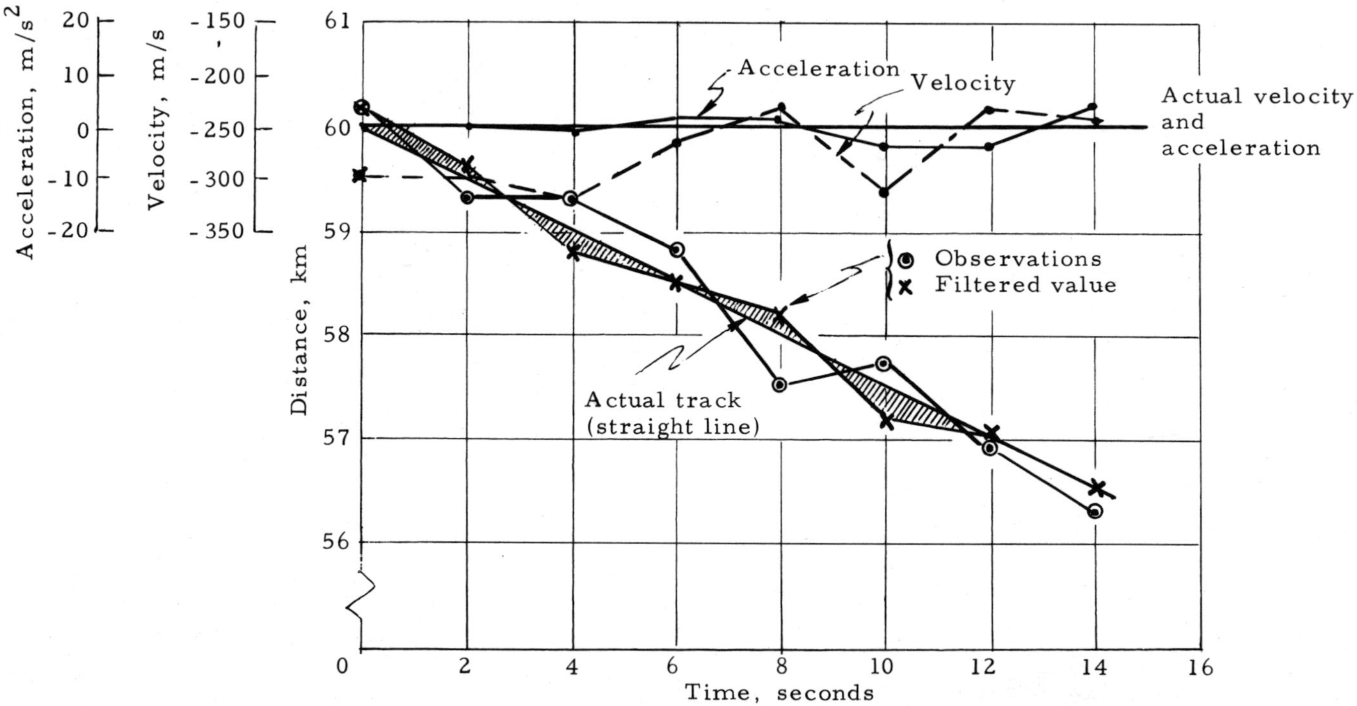

Figure 5-5. Actual and Filtered Values of Variables of Example Problem

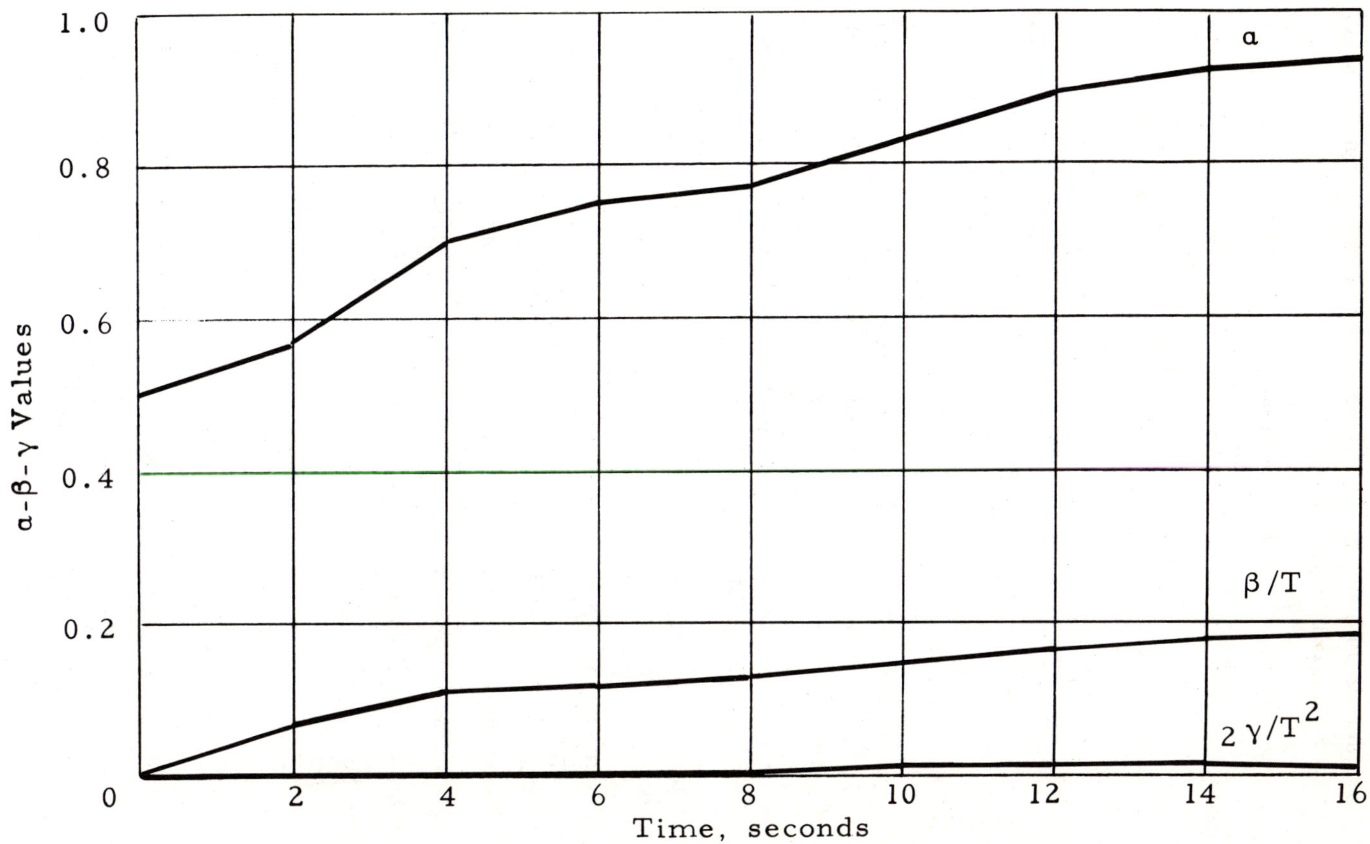

Figure 5-6. The Values of α-β-γ Computed for the Example Problem

4. Consider the following data obtained for the elevation angle of a given target by a ground-based radar:

Radar Scan	*Time, seconds*	*Angle, ϵ, degrees*
1	0	10.0
2	4	10.5
3	8	10.8
4	12	11.0
5	16	11.1
6	20	11.8
7	24	11.8
8	28	11.9
9	32	12.3
10	36	12.2

Write the α-β tracker equations for ϵ. Use a value of $\alpha = 0.3$, $\beta = 0.1$, and obtain the smooth and predicted values for a radar sampling time of 4 seconds. For the first time step, set $\epsilon_{S1} = \epsilon_{P2} = \epsilon_{O1}$ and $\dot{\epsilon}_1 = \ddot{\epsilon}_1 = 0$.

5. Same as problem 4, but this time use an α-β-γ tracker with $\alpha = 0.3$, $\beta = 0.1$, and γ and 0.05.

6. Obtain the values of $q_1 = q_2 = \sqrt{(1 - \alpha)}$ of equation (5.19) for the critical α-β tracker.

7. Obtain equation (5.41) using the cost function given in (5.38).

8. Obtain relation (5.49) describing the error covariance matrix.

9. Obtain the values of matrix equations (5.76) through (5.80).

10. Obtain the values of $\mathbf{K}_1$, $\hat{\mathbf{X}}_2$, and $\mathbf{P}_2$ of equation (5.82).

11. Obtain the values of $\mathbf{K}_2$ and $\hat{\mathbf{X}}_3$ of equations (5.83) and (5.84).

12. In the numerical example of Kalman filtering, assume the following values:

$$\mathbf{X}_0 = \begin{pmatrix} 40225 \\ -150 \\ 0 \end{pmatrix}; \quad \mathbf{Y}_0 = 40225$$

and

$$\mathbf{Y}_1 = 38960; \quad \mathbf{Y}_2 = 38975$$

With the rest of the values the same as given in the problem, calculate $\hat{\mathbf{X}}_1$ and $\mathbf{P}_1$.

13. In the numerical example of Kalman filtering, assume the following numerical values:

$$\mathbf{X}_0 = \begin{pmatrix} 38750 \\ -185 \\ 0 \end{pmatrix}; \quad \mathbf{Y}_0 = 38750$$

$$\mathbf{Y}_1 = 38300; \qquad \mathbf{Y}_2 = 38050$$

with the test of the values as given in the problem. Compute $\hat{\mathbf{X}}_2$, $\mathbf{K}_2$, and $\mathbf{P}_2$.

14. Consider the data of problem 4 with the additional information given below.

$$\phi = \begin{pmatrix} 1 & 4 & 8 \\ 0 & 1 & 4 \\ 0 & 0 & 1 \end{pmatrix}; \quad \mathbf{Q} = \begin{pmatrix} 0.01 & 0 & 0 \\ 0 & 0.0001 & 0 \\ 0 & 0 & 0.000001 \end{pmatrix}$$

$$\mathbf{M} = (1,0,0)\,; \ \mathbf{R} = 0.04\,; \quad \mathbf{P} = \begin{pmatrix} 0.04 & 0 & 0 \\ 0 & 0.0001 & 0 \\ 0 & 0 & 0.000001 \end{pmatrix}$$

$$\mathbf{Y}_0 = 10\,; \quad \mathbf{X}_0 = \begin{pmatrix} 10 \\ 0 \\ 0 \end{pmatrix}; \quad \mathbf{Y}_1 = 10.5\,; \quad \mathbf{Y}_2 = 10.8.$$

Calculate $\hat{\mathbf{X}}_1$, $\mathbf{K}_1$, $\mathbf{P}_1$ and $\hat{\mathbf{X}}_2$, $\mathbf{K}_2$, $\mathbf{P}_2$.

15. Consider a system where R, $\dot{R}$, and $\ddot{R}$ are measured every 10 seconds. Formulate this system as a Kalman filtering problem, using the following matrices:

$$\mathbf{X} = \begin{pmatrix} R \\ \dot{R} \\ \ddot{R} \end{pmatrix}; \quad \phi = \begin{pmatrix} 1 & 10 & 50 \\ 0 & 1 & 10 \\ 0 & 0 & 1 \end{pmatrix}; \quad \mathbf{Q} = \begin{pmatrix} 400 & 0 & 0 \\ 0 & 100 & 0 \\ 0 & 0 & 25 \end{pmatrix};$$

$$\mathbf{M} = \begin{pmatrix} 1 & 0 & 0 \\ 0 & 1 & 0 \\ 0 & 0 & 1 \end{pmatrix}; \quad \mathbf{R} = \begin{pmatrix} 40000 & 0 & 0 \\ 0 & 400 & 0 \\ 0 & 0 & 10 \end{pmatrix};$$

$$\mathbf{P}_0 = \begin{pmatrix} 40000 & 0 & 0 \\ 0 & 2500 & 0 \\ 0 & 0 & 400 \end{pmatrix}; \quad \mathbf{X}_0 = \begin{pmatrix} 50000 & 0 & 0 \\ -200 & 0 & 0 \\ 10 & 0 & 0 \end{pmatrix};$$

$$\mathbf{Y}_0 = \begin{pmatrix} 50000 & 0 & 0 \\ -200 & 0 & 0 \\ 10 & 0 & 0 \end{pmatrix}; \quad \mathbf{Y}_1 = \begin{pmatrix} 48125 & 0 & 0 \\ -210 & 0 & 0 \\ 5 & 0 & 0 \end{pmatrix};$$

$$\mathbf{Y}_2 = \begin{pmatrix} 45500 & 0 & 0 \\ -175 & 0 & 0 \\ 15 & 0 & 0 \end{pmatrix}$$

Using the above data, calculate $\mathbf{P}_1$, $\hat{\mathbf{X}}_1$ and $\mathbf{P}_2$, $\hat{\mathbf{X}}_2$.

Appendix 5A Recursive Relations of α-β Trackers

Consider the α-β tracker equations given in the text and repeated here for convenience

$$\hat{x}_n = \alpha x_n + (1 - \alpha)x_{p_n} \tag{5.A1}$$

$$\dot{x}_n = \dot{x}_{n-1} + \frac{\beta}{T}(x_n - x_{p_n}) \tag{5.A2}$$

$$x_{p(n+1)} = x_n + T\dot{x}_n \tag{5.A3}$$

Solving equation (5.A3) for $\dot{x}_n$, we get

$$\dot{x}_n = \frac{1}{T}(x_{p(n+1)} - \hat{x}_n) \tag{5.A4}$$

Replacing n by $n - 1$ in the above results in

$$\dot{x}_{n-1} = \frac{1}{T}(x_{p_n} - \hat{x}_{n-1}) \tag{5.A5}$$

Substituting (5.A4) and (5.A5) in (5.A2), we get

$$x_{p(n+1)} - \hat{x}_n = x_{p_n} - \hat{x}_{n-1} + \beta x_n - \beta x_{p_n} \tag{5.A6}$$

Solve equation (5.A1) for x_{p_n} as follows:

$$x_{p_n} = \frac{1}{1 - \alpha}(\hat{x}_n - \alpha x_n) \tag{5.A7}$$

Replace n by $n + 1$ and get

$$x_{p(n+1)} = \frac{1}{1 - \alpha}(\hat{x}_{n+1} - \alpha x_{n+1}) \tag{5.A8}$$

Substituting (5.A7) and (5.A8) in equation (5.A6) and simplifying, we get

$$\hat{x}_{n+1} - \alpha x_{n+1} - (1 - \alpha)\hat{x}_n = \hat{x}_n - \alpha x_n - (1 - \alpha)\hat{x}_{n-1} + \beta(1 - \alpha)x_n - \beta\hat{x}_n + \alpha\beta x_n \tag{5.A9}$$

Replacing n with $n - 1$ and solving for $\hat{x}_n$, we get

$$\hat{x}_n = (2 - \alpha - \beta)\hat{x}_{n-1} - (1 - \alpha)\hat{x}_{n-2} + \alpha x_n + (\beta - \alpha)x_{n-1} \tag{5.A10}$$

Equation (5.A10) is the same as equation (5.14) of the text.

Notes

1. T.R. Benedict and G.W. Bordner, "Synthesis of an Optimal Set of Radar Track-While-Scan Smoothing Equations," *IRE Transactions on Automatic Control* (July 1962), pp. 27-32.

2. C. Jordan, *Calculus of Finite Differences*, (New York: Chelsea Publishing Company, 1947), chap. 2.

3. Benedict and Bordner, "Synthesis of an Optimal Set," pp. 27-32.

4. Ibid.

5. H.R. Simpson, "Performance Measures and Optimization Conditions for a Third-Order Sample-Data Tracker," *IRE Transactions on Automatic Control* (April 1963), pp. 182-183.

6. Ibid.

7. S.A. Hovanessian and Louis A. Pipes, *Digital Computer Methods in Engineering* (New York: McGraw-Hill, 1969), p. 262.

Bibliography

Leondes, C.T. (Editor) *Advances in Control Systems* Vol. 3, New York: Academic Press, 1966, pp. 219-289, "Kalman Filtering Techniques," by H.W. Sorenson.

Luenberger, D.G. *Optimization by Vector Space Method*. New York: Wiley 1969, Chapter 4.

Nahi, N.E. *Estimation Theory and Applications*. New York: Wiley, 1969.

Taylor, F.J. "On the Computation of Kalman Gains," *Computers and Electrical Engineering*. London: Pergamon Press, 1975, pp. 105-115.

6 Spectral Analysis and the Fast Fourier Transform

In engineering analyses the parameters of system behavior are either expressed as a function of time or as a function of frequency contents. For example, the solution of differential equations of motion of rigid bodies will result in the values of displacements, velocities, and accelerations as a function of time, while oscillations of electrical and mechanical systems and doppler shifts of radar return are expressed in the frequency domain. Time and frequency domains are related through Fourier series and transforms.

Given a variable expressed as a function of time, Fourier series will decompose the variable into a sum of oscillatory functions, each having a specific frequency. These frequencies, with their corresponding amplitudes and phase angles, constitute the frequency contents of the original time functions.

This chapter will start from Fourier series[1] and develop the Fourier transform together with several newly developed computational schemes. It will discuss the relationship of Fourier analysis and digital filtering methods together with an example taken from industry.

Fourier Series

Theorem: Let $f(t)$ be a function defined in the interval $-\pi \leq t \leq \pi$ and outside this interval $f(t + 2\pi) = f(t)$, then $f(t)$ can be represented by a series

$$f(t) = \frac{a_0}{2} + \sum_{k=1}^{\infty} (a_k \cos kt + b_k \sin kt) \qquad (6.1)$$

with

$$a_k = \frac{1}{\pi} \int_{-\pi}^{\pi} f(t) \cos kt \, dt \qquad (6.2)$$

$$b_k = \frac{1}{\pi} \int_{-\pi}^{\pi} f(t) \sin kt \, dt \qquad k = 0, 1, 2, \ldots$$

which converges at every point t_0 to

$$f(t_0) = \frac{f(t_0 +) + f(t_0 -)}{2} \tag{6.3}$$

Equation (6.3) at a point of discontinuity will give the average value of the function $f(t)$, as seen in the following example.

Example of Fourier Series

Consider the function

$$\begin{aligned} f(t) &= 0 \qquad -\pi < t < 0 \\ &= \pi \qquad 0 < t < \pi \end{aligned} \tag{6.4}$$

The Fourier coefficients can be calculated as

$$\begin{aligned} a_0 &= \pi \\ a_n &= 0 \\ b_n &= \frac{1}{n}(1 - \cos n\pi) \end{aligned} \tag{6.5}$$

The series then can be written as

$$f(t) = \frac{\pi}{2} + 2\left(\frac{\sin t}{1} + \frac{\sin 3t}{3} + \frac{\sin 5t}{5} + \dots\right) \tag{6.6}$$

The graph of this series is given in Figure 6.1. Note the values at discontinuities $-\pi$, 0, and π, as obtained from the series and compare with the corresponding values from equation (6.3).

Sine and Cosine Series

If $f(x)$ is an *even* function, so that

$$f(x) = f(-x) \tag{6.7}$$

Fourier series will consist only of cosine terms.

If $f(x)$ is an *odd* function, so that

$$f(x) = -f(-x) \tag{6.8}$$

then the Fourier series will consist only of sine terms.

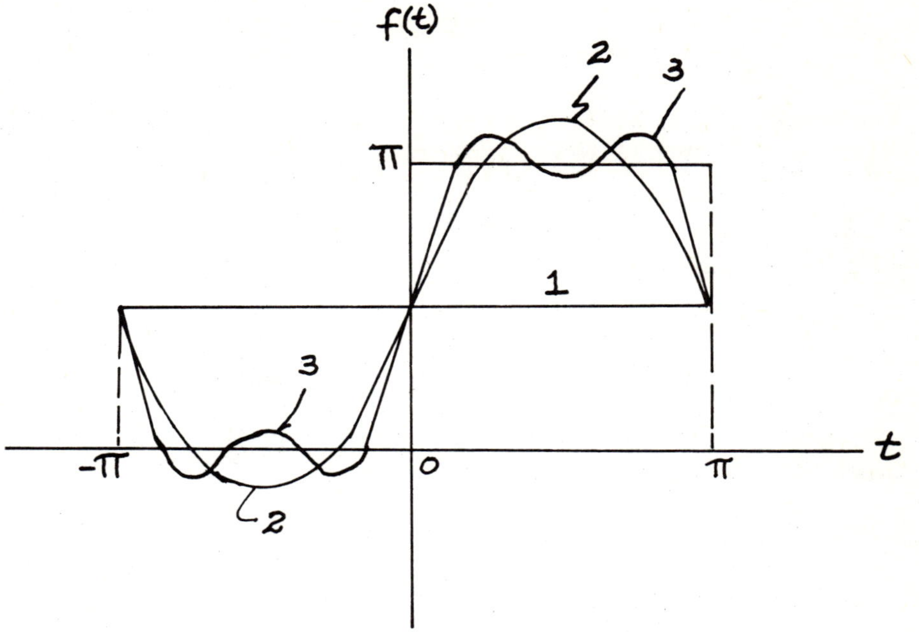

Figure 6-1. Example of Fourier Series (Numbers refer to sum of terms)

Convergence of Fourier Series

Theorem: If the function $f(x)$ is periodic and has $k - 1$ derivatives in the interval $-\pi < x < \pi$, and such that its kth derivative satisfies the Dirichlet condition (this means that the function has a finite number of discontinuities in the interval $-\pi < x < \pi$), the Fourier coefficients satisfy the inequalities

$$|a_n| < M/n^{k+1} \qquad |b_n| < M/n^{k+1} \tag{6.9}$$

where brackets represent absolute values and M is a positive number independent of the nth term of the series. Consider these functions together with their convergence criteria. In Figure 6-2, $f(x)$ itself is discontinuous. This results in a convergence rate of $1/n$. This case was illustrated by the examples in Figure 6-1, where the series of equation (6.6) converged with a rate of $1/n$. In Figure 6-3, the function $f(x)$ is continuous, while its derivative with respect to $x, f'(x)$ is discontinuous. The convergence rate of the Fourier series will be proportional to $1/n^2$. This can be seen by evaluating the Fourier coefficients and writing the series for $f(x)$:

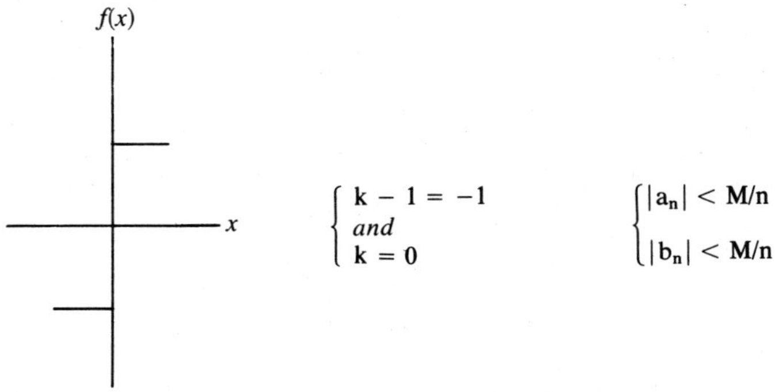

Figure 6-2. Discontinuous Function $f(x)$ vs. x

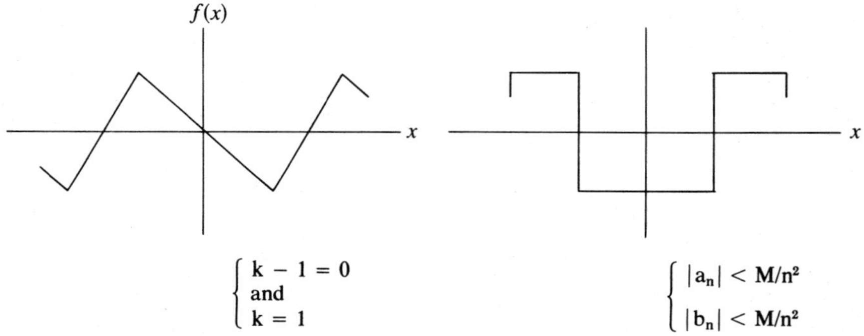

Figure 6-3. Continuous Function $f(x)$ with Discontinuous Derivative $f'(x)$

$$f(x) = \frac{A}{\pi}\left(\sin \omega t - \frac{1}{3^2} \sin 3 \omega t + \frac{1}{5^2} \sin 5 \omega t - \dots\right) \qquad (6.10)$$

It is noted that this series converges as $1/n^2$ in agreement with the previous discussion.

Complex Fourier Series and Transforms

We can derive the Fourier integral formulas by starting from the complex form of the Fourier series. This form of the Fourier series, in turn, can be derived from the trigonometric form by using the relation

$$e^{j\sigma} = \cos \sigma + j \sin \sigma \qquad (6.11)$$

where $j = \sqrt{(-1)}$. The complex form of the Fourier series is written

$$F(t) = \sum_{-\infty}^{\infty} a_n e^{jn\omega t} \qquad T = \frac{2\pi}{\omega} \qquad (6.12)$$

where summation is over the index n, and

$$a_n = \frac{1}{T}\int_{-T/2}^{T/2} F(u)e^{-jn\omega u}\,du \tag{6.13}$$

Substituting a_n from (6.13) in (6.12), we get

$$F(t) = \sum_{-\infty}^{\infty}\frac{1}{T}\int_{-T/2}^{T/2} F(u)e^{(2\pi nj/T)(t-u)}\,du \tag{6.14}$$

Let us set $1/T = \Delta S$

$$F(t) = \sum_{-\infty}^{\infty}\Delta S\int_{-T/2}^{T/2} F(u)e^{2\pi nj(t-u)\,\Delta S}\,du \tag{6.15}$$

Consider the following equation which relates the integration of $\phi(S)$ to summation over discrete intervals ΔS

$$\int_{0}^{\infty}\phi(S)\,dS = \lim_{\Delta S\to 0}\sum_{n=0}^{\infty}\phi(n\Delta S)\,\Delta S \tag{6.16}$$

Extending the limits of the integral from $-\infty$ to ∞, we get

$$\int_{-\infty}^{\infty}\phi(S)\,dS = \lim_{\Delta S\to 0}\sum_{-\infty}^{\infty}\phi(n\Delta S)\,\Delta S \tag{6.17}$$

As $T \to \infty$, we get from (6.15)

$$F(t) = \int_{-\infty}^{\infty} e^{2\pi jSt}\left[\int_{-\infty}^{\infty} F(u)\,e^{-2\pi jSu}\,du\right]ds \tag{6.18}$$

Introducing the variable $\omega = 2\pi S$ in (6.18), we get

$$F(t) = \frac{1}{2\pi}\int_{-\infty}^{\infty} e^{j\omega t}\left[\int_{-\infty}^{\infty} F(u)\,e^{-j\omega u}\,du\right]d\omega \tag{6.19}$$

Let us write equation (6.19) in the form

$$F(t) = \int_{-\infty}^{\infty} G(\omega)\,e^{j\omega t}\,d\omega \tag{6.20}$$

$$G(\omega) = \frac{1}{2\pi}\int_{-\infty}^{\infty} F(t)\,e^{-j\omega t}\,dt \tag{6.21}$$

These equations represent Fourier transform pairs. Equation (6.21) represents the Fourier transform of $F(t)$, while (6.20) represents the inverse transform.

Example: Find the Fourier transform of pulse $F(t)$ given in Figure 6-4. Using equation (6.21), we get

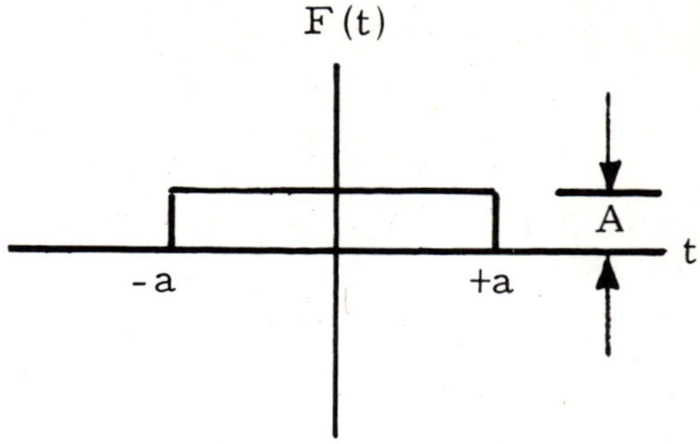

Figure 6-4. *F*(*t*) for the Example Problem

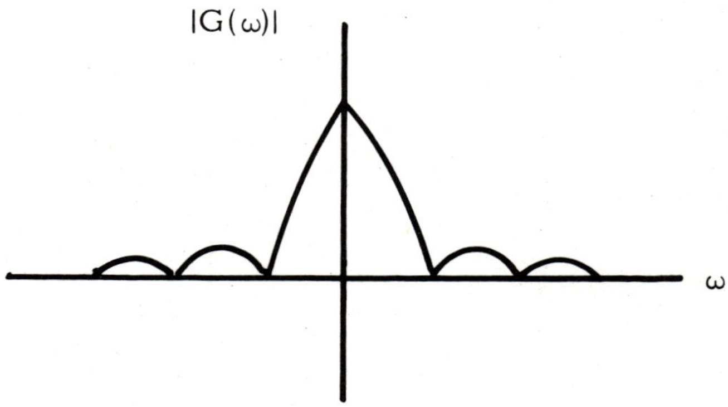

Figure 6-5. Solution of the Example Problem

$$\begin{aligned} G(\omega) &= \frac{A}{2\pi}\int_{-a}^{a} e^{-j\omega t}\,dt \\ &= \frac{A}{2\pi}\int_{-a}^{a} (\cos \omega t - j \sin \omega t)\,dt \\ &= \frac{A}{2\pi}\left[\frac{\sin \omega t}{\omega} + j\,\frac{\cos \omega t}{\omega}\right]_{-a}^{a} \\ &= \frac{A \sin \omega a}{\pi\omega} \end{aligned} \tag{6.22}$$

Figure 6.5 gives the sin x/x type solution of the problem.

Numerical Evaluation of Fourier Transform[2]

Consider the representation of $F(t)$ and $G(\omega)$ from Figure 6-6. The following notation applies

$$\begin{aligned} T &= \text{total time interval} \\ K &= \text{number of time increments} \\ \Delta t &= \text{time increment} \\ N &= \text{number of frequency increments} \\ \Delta\omega &= \text{frequency increment} \end{aligned} \tag{6.23}$$

To simplify the following calculations let us denote

$$K = 2^{\gamma} \quad \text{and} \quad N = 2^{\gamma+\gamma_1} \tag{6.24}$$

where γ and γ_1 are integers. The frequency interval is defined below

$$\Delta\omega = 2\pi\Delta f = 2\pi \frac{1}{T \cdot N/K} = \frac{2\pi K}{NT} \tag{6.25}$$

Using this notation in (6.20, 21), the Fourier transform pair can be written in discrete form

$$F(t_k) = \Delta\omega \sum_{n=0}^{N-1} G(\omega_n)\exp(j\omega_n t_k) \quad k = 0, \ldots, K - 1 \tag{6.26}$$

$$G(\omega_n) = \frac{\Delta t}{2\pi} \sum_{k=0}^{K-1} F(t_k)\exp(-j\omega_n t_k) \quad n = 0, \ldots, N - 1 \tag{6.27}$$

Using the following notation

$$\begin{aligned} F_k &= \frac{T}{2\pi K} F(t_k) \\ G_n &= G(\omega_n) \\ \omega_n &= n\Delta\omega = \frac{2\pi nK}{NT} \\ t_k &= k\Delta t = \frac{kT}{K} \end{aligned} \tag{6.28}$$

we can write equations (6.26) and (6.27) as

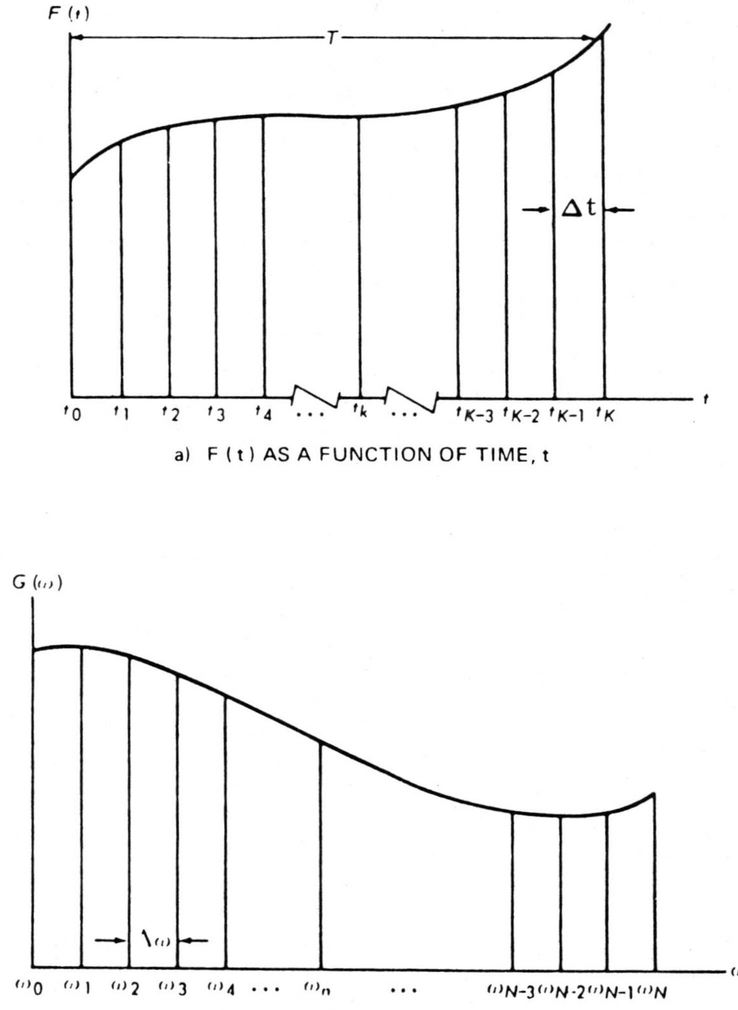

a) F (t) AS A FUNCTION OF TIME, t

b) G (ω) AS A FUNCTION OF ANGULAR FREQUENCY, ω

Figure 6-6. Discrete Representation of $F(t)$ and $G(\omega)$

$$F_k = \frac{1}{N} \sum_{n=0}^{N-1} G_n \exp[(2\pi j/N)(nk)] \qquad k = 0, \ldots, K-1 \tag{6.29}$$

$$G_n = \sum_{k=0}^{K-1} F_k \exp[(-2\pi j/N)(nk)] \qquad n = 0, \ldots, N-1 \tag{6.30}$$

Denoting $W = \exp(-2\pi j/N)$, we represent equation (6.29, 30) as a matrix product as follows:

$$\underset{(K\times 1)}{\mathbf{F}_k} = \frac{1}{N} \underset{(K\times N)}{\mathbf{W}^{-kn}} \underset{(N\times 1)}{\mathbf{G}_n} \tag{6.31}$$

$$\underset{(N\times 1)}{\mathbf{G}_n} = \underset{(N\times K)}{\mathbf{W}^{nk}} \underset{(K\times 1)}{\mathbf{F}_k} \tag{6.32}$$

where product nk, in addition to being the power of W is also used to give the position of W^{nk} term in the matrix. Equation (6.31) represents frequency-to-time-domain transformation (inverse Fourier transform) and equation (6.32) represents time-to-frequency-domain transformation (Fourier transform).

Computation of Powers of W

Consider W raised to the power l

$$\begin{aligned} W^l &= \exp[(-2\pi j/N)l] \\ &= \cos\,(2\pi/N)l - j\sin\,(2\pi/N)l \end{aligned} \tag{6.33}$$

where $l = nk$ can be computed as follows

$$\begin{aligned} l &= nk \text{ modulus of } N \\ &= \text{integer remainder after division of } nk \text{ by } N \end{aligned} \tag{6.34}$$

Under these conditions we need to compute W to the powers $0, 1, \ldots, N-1$ instead of (max n) (max k) or $(K-1)(N-1)$. This allows a considerable computational saving.

As an example, consider W to the powers 0 through 4 as given in Figure 6-7 and described below:

$$\begin{aligned} W^l &= \exp[(-2\pi j/4)l] \\ &= \cos\left(\frac{\pi l}{2}\right) - j\sin\left(\frac{\pi l}{2}\right) \end{aligned} \tag{6.35}$$

$$W^0 = 1,\ W^1 = -j,\ W^2 = -1,\ W^3 = j,\ W^4 = 1$$

Note the symmetry of W values with respect to the real and the imaginary axis of Figure 6-7.

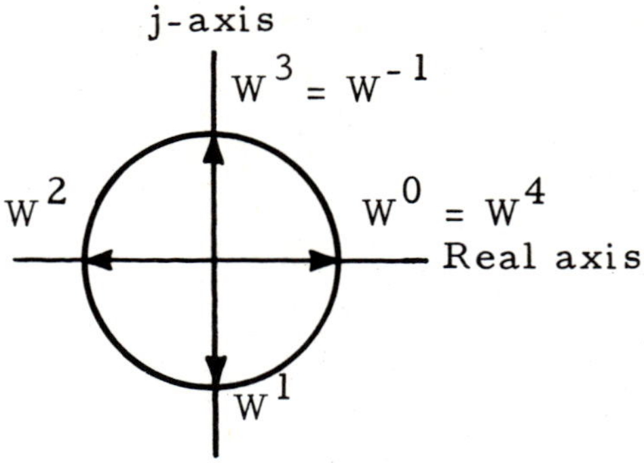

Figure 6-7. Powers of W with $N = 4$

Matrix Computation of Fourier Transform

Consider the matrix representation of equation (6.32) for $N = K = 4$

$$\begin{pmatrix} G_0 \\ G_1 \\ G_2 \\ G_3 \end{pmatrix} = \begin{pmatrix} W^{0(0,0)} & W^{0(0,1)} & W^{0(0,2)} & W^{0(0,3)} \\ W^{0(1,0)} & W^{1(1,1)} & W^{2(1,2)} & W^{3(1,3)} \\ W^{0(2,0)} & W^{2(2,1)} & W^{4(2,2)} & W^{6(2,3)} \\ W^{0(3,0)} & W^{3(3,1)} & W^{6(3,2)} & W^{9(3,3)} \end{pmatrix} \begin{pmatrix} F_0 \\ F_1 \\ F_2 \\ F_3 \end{pmatrix} \tag{6.36}$$

And, using $l = nk$ modulus of N in the powers of W, we get

$$\begin{pmatrix} G_0 \\ G_1 \\ G_2 \\ G_3 \end{pmatrix} = \begin{pmatrix} W^0 & W^0 & W^0 & W^0 \\ W^0 & W^1 & W^2 & W^3 \\ W^0 & W^2 & W^0 & W^2 \\ W^0 & W^3 & W^2 & W^1 \end{pmatrix} \begin{pmatrix} F_0 \\ F_1 \\ F_2 \\ F_3 \end{pmatrix} \tag{6.37}$$

Using the standard methods of multiplication to carry out the multiplication of equation (6.37) will require 16 multiplications, but note the symmetrical conditions depicted below:

$$\begin{array}{llllllll} G_0 = F_0 + & & W^0F_1 & & + W^0F_2 & & +W^0F_3 & \\ G_1 = F_0 + & \text{complex} & -W^1F_1 & & + W^2F_2 & \text{complex} & +W^3F_3 & \\ & & & \text{Repeat} & & & & (6.38) \\ G_2 = F_0 + & \text{conjugate} & W^2F_1 & & + W^0F_2 & \text{conjugate} & +W^2F_3 & \\ G_3 = F_0 + & & W^3F_1 & & W^2F_2 & & +W^1F_3 & \end{array}$$

Noting the conditions in (6.38), we may apply the following rules to save multiplications:

1. Obtain the values of W^l, $l = 0, 1, \ldots, N - 1$.
2. Set $G_0 \ldots G_{N-1}$ equal to F_0.
3. Compute $l = nk$ modulus N, starting with $k = 1$ and $n = 0, \ldots, N - 1$ multiply W by F_k. When $l = N/2$, obtain the complex conjugate of previously computed WF products, symmetrically located with respect to $W^{N/2}$. If $l = 0$, the rest of WF values repeat.
4. Add the values of step (3) to $G_0, G_1, \ldots G_{N-1}$ values.
5. Repeat 3 and 4 for $k = 2, \ldots, N - 1$.

The resulting values of Gs will have the form

$$G_n = a_n + jb_n \tag{6.39}$$

or

$$\begin{aligned} G_n &= |G_n| (\cos \psi_n + j \sin \psi_n) \\ &= |G_n| \, e^{j\psi_n} \end{aligned} \tag{6.40}$$

where

$$\begin{aligned} |G_n| &= \text{absolute value of } G_n \\ &= \sqrt{a_n^2 + b_n^2} \\ \psi_n &= \text{arc tan}(b_n/a_n) \end{aligned} \tag{6.41}$$

Example 1: As the first example of the procedure in (6.38), consider the spectral representation of the function of Figure 6-8. The values of $K = N = 16$ are selected, resulting in $\Delta\omega = 2\pi/16 = 0.3927$.

Table 6-1 gives the input data together with the resulting output using a computer program. It is noted that the foldover occurs at $\omega = \pi$ where the values of G are complex conjugate of previously computed values.

Having obtained the values of Gs, we may use the inversion integral of equation (6.26) and compute the values of $F(t)$. These, of course, will equal the input data as given in Table 6-2.

Cooley-Tukey Method

This method is based on expressing the **W** matrix of equation (6.32) in terms of γ square matrices where γ is obtained from $K = N = 2^\gamma$.

Consider equation (6.37) with $N = 4$.

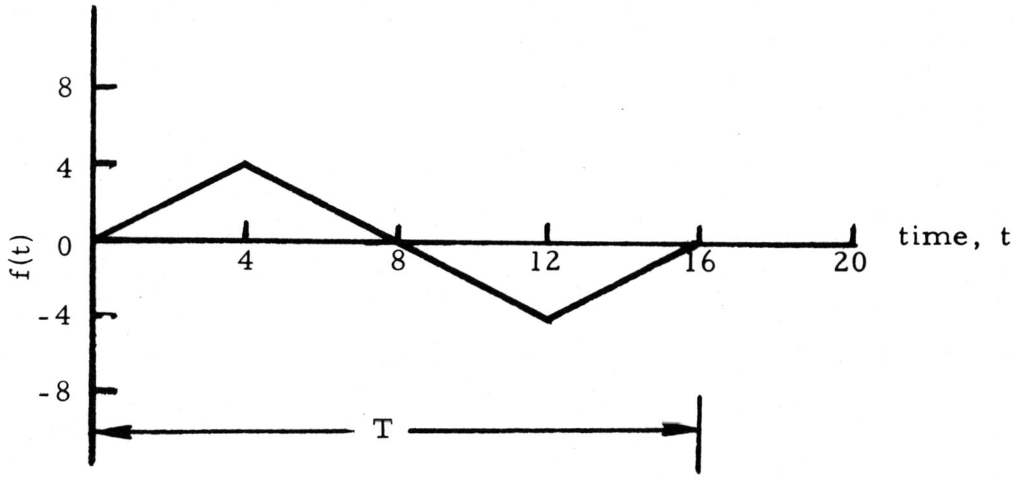

Figure 6-8. The Function $f(t)$ vs. Time, Example 1

Table 6-1
Solution of the Example Problem No. 1

Time step k	Time t	Function $F(t)$
0	0	0
1	1	1
2	2	2
3	3	3
4	4	4
5	5	3
6	6	2
7	7	1
8	8	0
9	9	−1
10	10	−2
11	11	−3
12	12	−4
13	13	−3
14	14	−2
15	15	−1

Freq. step n	Omega ω	Absolute value of $\lvert G \rvert$	Real part of G	Imaginary part of G
0	0	0	0	0
1	0.392699	4.18166	0	−4.18166
2	0.785398	0	0	0
3	1.1781	0.515635	0	0.515635
4	1.5708	0	0	0
5	1.9635	0.230212	0	−0.230212
6	2.35619	0	0	0
7	2.74889	0.165452	0	0.165452
8	3.14159	0	0	0
9	3.53429	0.165452	0	−0.165452
10	3.92699	0	0	0
11	4.31969	0.230212	0	0.230212
12	4.71239	0	0	0
13	5.10509	0.515635	0	0.515635
14	5.49779	0	0	0
15	5.89049	4.18166	0	4.18166

$$\begin{pmatrix} G_0 \\ G_1 \\ G_2 \\ G_3 \end{pmatrix} = \begin{pmatrix} W^0 & W^0 & W^0 & W^0 \\ W^0 & W^1 & W^2 & W^3 \\ W^0 & W^2 & W^0 & W^2 \\ W^0 & W^3 & W^2 & W^1 \end{pmatrix} \begin{pmatrix} F_0 \\ F_1 \\ F_2 \\ F_3 \end{pmatrix} \tag{6.42}$$

Noting that $W^0 = 1$, we can write

Table 6-2
The Inverse Transform

Time Step	*Real Part*	*Imaginary Part*
0	0.	0
1	1.	0
2	2.	0
3	3.	0
4	4.	0
5	3.	0
6	2.	0
7	1.	0
8	0	0
9	−1.	0
10	−2.	0
11	−3.	0
12	−4.	0
13	−3.	0
14	−2.	0
15	−1.	0

$$\begin{pmatrix} G_0 \\ G_2 \\ G_1 \\ G_3 \end{pmatrix} = \begin{pmatrix} 1 & W^0 & 0 & 0 \\ 1 & W^2 & 0 & 0 \\ 0 & 0 & 1 & W^1 \\ 0 & 0 & 1 & W^3 \end{pmatrix} \begin{pmatrix} 1 & 0 & W^0 & 0 \\ 0 & 1 & 0 & W^0 \\ 1 & 0 & W^2 & 0 \\ 0 & 1 & 0 & W^2 \end{pmatrix} \begin{pmatrix} F_0 \\ F_1 \\ F_2 \\ F_3 \end{pmatrix} \tag{6.43}$$

Note the change in the order of the **G** matrix. The product of the last two matrices is

$$\mathbf{F}^1 = \begin{pmatrix} F_0^1 \\ F_1^1 \\ F_2^1 \\ F_3^1 \end{pmatrix} = \begin{pmatrix} F_0 + W^0F_2 \\ F_1 + W^0F_3 \\ F_0 + W^2F_2 \\ F_1 + W^2F_3 \end{pmatrix} \tag{6.44}$$

Having the $\mathbf{F}^1$ matrix, we can compute the elements of the **G** matrix of equation (6.43).

The matrix multiplication of (6.43) can also be accomplished by using the computation tree in Figure 6-9.

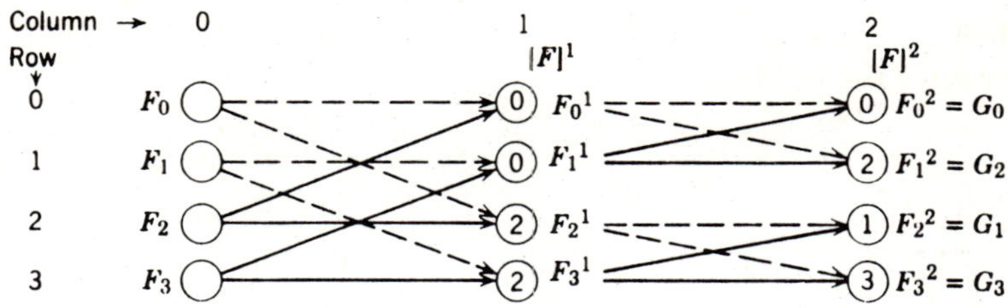

Figure 6-9. Computation Tree for $N = 4$

Solid lines represent multiplications and dashed lines are addition in Figure 6-9. The numbers inside the circle are powers of W. For example:

$$F_0^1 = F_0 + W^0 F_2$$

$$F_1^1 = F_1 + W^0 F_3$$

$$F_2^1 = F_0 + W^0 F_2$$

$$F_3^1 = F_0 + W^0 F_3 \tag{6.45}$$

Also

$$F_0^2 = G_0 = F_0^1 + F_1^1 W^0$$

$$F_1^2 = G_2 = F_0^1 + F_1^1 W^2 \tag{6.46}$$

$$\cdots$$

which can be verified from equation (6.43).

Rule for the Order of **G** *Matrix*

The order of the **G** matrix of (6.43) is obtained in Tables 6-3 and 6-4. The last column in Table 6-3 gives the subscripts of the **G** matrix as given in equation (6.43) and Figure 6-9.

Rules for the Computation Tree of Figure 6-9

1. Assume N sampled values of $F(t)$ where $N = 2^\gamma$ (γ is the number of binary bits in N).
2. Arrange γ columns for $[F]^1, [F]^2, \ldots, [F]^\gamma$.

Table 6-3
The Order of the G Matrix
($N = 4$)

Original Subscript	*Binary from γ Bits*	*Symmetrically Invested*	*Final Subscript*
0	00	00	0
1	01	10	2
2	10	01	1
3	11	11	3

Table 6-4
The Order of the G Matrix
($N = 8$)

Original Subscript	*Binary from 3 Bits*	*Symmtrically Invested*	*Final Subscript*
0	000	000	0
1	001	100	4
2	010	010	2
3	011	110	6
4	100	001	1
5	101	101	5
6	110	011	3
7	111	111	7

3. Let the numbers in the circles representing powers of W be denoted by P. For row k and column l (starting from k and l equal to zero):
   a. Write the binary number k.
   b. Slide this number $\gamma - l$ places to the right, filling newly opened places with zeros.
   c. Reverse the order of resulting binary bits. The integer is P.

   For $N = 8 = 2^3$, the power in the 4th row ($k = 4$) and second column ($l = 2$) is

$$\gamma - l = 3 - 2 = 1$$

a. $k = 4 =$ 100
b. 010
c. 010 P = 2

4. For multiplication and division, denote the row number k by binary bits,

$$k = k_{\gamma-1} \ldots k_2 k_1 k_0$$

In column l, node k has a solid line drawn to it from a node in the $(l - 1)$ column, with the bit location $k_{\gamma-l}$ replaced by one.

For $N = 16 = 2^4 (\gamma = 4)$, the element of the third column $(l = 3)$ fifth row $(k = 5)$ will have a solid line from the element of the second column having the row number $k = k_3 k_2 k_1 k_0 = 5 = 0101$ with $k_{\gamma-l} = k_{4-3} = k_1$ replaced by 1,

$$k = 0111 = 7 \tag{6.47}$$

(solid line from column 2 and row 7).

The dashed line comes from a node in $(l - 1)$ column with binary representation of k, where $k_{\gamma-l}$ is replaced by zero. In the above example, this will result in $k = 5$.

Using the Cooley-Tukey logic, special-purpose computers using binary arithmetic are built to carry out the calculations.

The number of multiplications in the Cooley-Tukey method are $N \log_2 N$. These are complex multiplications, each representing four multiplications of real numbers. This should be compared to N^2 multiplication if direct matrix multiplication of equation (6.32) is employed.

Analytic Derivation of Cooley-Tukey Relations

Consider again the matrices of equation (6.32) written below in the summation form

$$G_n = \sum_{k=0}^{K-1} F_k W^{nk} \tag{6.48}$$

where G_n represents the Fourier transform and F_k is the sampled data points and $W = \exp(-2\pi j/N)$. We note that for $N = K$ we have $n = 0, 1, \ldots, N - 1$ and $k = 0, 1, \ldots, N - 1$. To make the derivation of equations more specific, let us set $N = 8$ and write the binary representation for n and k

$$n = n_2 4 + n_1 2 + n_0\,; \quad k = k_2 4 + k_1 2 + k_0 \tag{6.49}$$

Using (6.48) and substituting (6.49), we get

$$G_{n_2,n_1,n_0} = \sum_{k_0=0}^{1} \sum_{k_1=0}^{1} \sum_{k_2=0}^{1} F_{k_2,k_1,k_0} W^{(n_2 4+n_1 2+n_0)(k_2 4+k_1 2+k_0)} \tag{6.50}$$

$$G_{n_2,n_1,n_0} = \sum_{k_0=0}^{1} \sum_{k_1=0}^{1} \sum_{k_2=0}^{1} F_{k_2,k_1,k_0} W^{n_0 k_2 4} W^{(n_1 2+n_0)k_1 2} W^{(n_2 4+n_1 2+n_0)k_0}$$

$A_1(n_0 k_1 k_0)$

$A_2(n_0, n_1, k_0)$

$A_3(n_0, n_1, n_2)$

Figure 6-10. Relation Representing the Result of Cooley-Tukey Computation ($N = 8$)

Noting $W^{a+b} = W^a W^b$, we can write the W part of (6.32) as

$$W^{(n_2 4+n_1 2+n_0)(k_2 4+k_1 2+k_0)} = W^{(n_2 4+n_1 2+n_0)k_2 4} \cdot W^{(n_2 4+n_1 2+n_0)k_1 2} \cdot W^{(n_2 4+n_1 2+n_0)k_0} \tag{6.51}$$

We may group each of the right side terms as follows:

$$W^{(n_2 4+n_1 2+n_0)k_2 4} = [W^{8(n_2 2+n_1)k_2}] W^{(n_0 k_2 4)}$$

$$W^{(n_2 4+n_1 2+n_0)k_1 2} = [W^{8(n_2 k_1)}] W^{(n_1 2+n_0)k_1 2}$$

$$W^{(n_2 4+n_1 2+n_0)k_0} = W^{(n_2 4+n_1 2+n_0)k_0} \tag{6.52}$$

Using the relation

$$W^8 = [e^{-2\pi j/8}]^8 = 1 \tag{6.53}$$

in the first two of equation (6.52) we get, after substitution in equation (6.50), the relation in Figure 6-10.
From Figure 6-10 we get

$$G_{(n_2,n_1,n_0)} = A_3(n_0, n_1, n_2) \tag{6.54}$$

Note that the subscripts of G in equation (6.54) are symmetrically inverted as compared to indices of A_3, as discussed before.

Aliasing Effect

The aliasing effect refers to the fact that high-frequency components of a time function can be misrepresented as low frequencies if the sampling rate is too low. Figure 6-11 shows this effect. To avoid aliasing effects on

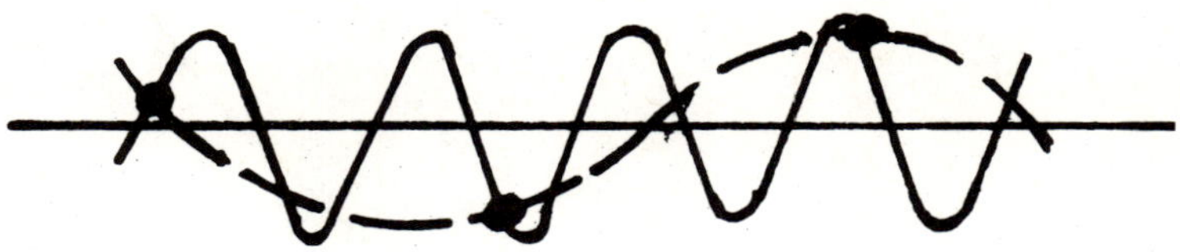

Figure 6-11. Misrepresentation of a High-frequency Wave Form by a Low-Frequency Sampling

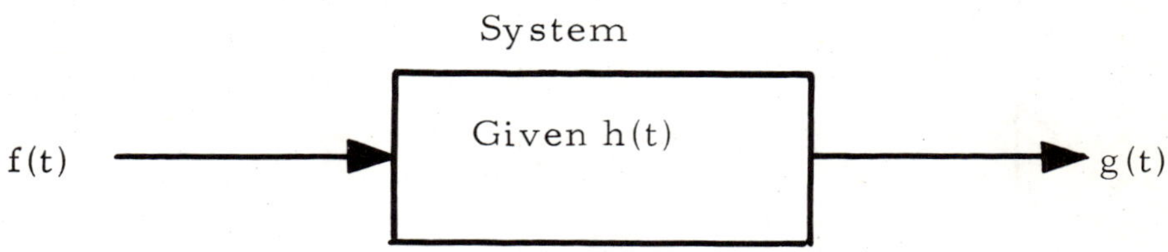

Figure 6-12. System Response to Input $f(t)$

a sampled-time function, the sampling frequency should be at least twice as high as the highest frequency present.

Digital Filtering

This section intends to show that, given the response of the system $h(t)$ to an impulse function, one can obtain the response of the system $g(t)$ to any given function $f(t)$ using the convolution integral (see Figure 6-12). Having obtained the time response of the system $g(t)$, the Fourier transform of $g(t)$ will give us the frequency response of the system. Calculations dealing with frequency response of systems using Fourier transforms and digital computers come under the general category of *digital filtering*.

Mathematically, the convolution integral is written

$$g(t) = \int_0^t f(\tau)h(t - \tau)\,d\tau \tag{6.55}$$

Note that the convolution integral operates in the time domain.

The convolution integral is evaluated numerically by the following consideration: Let

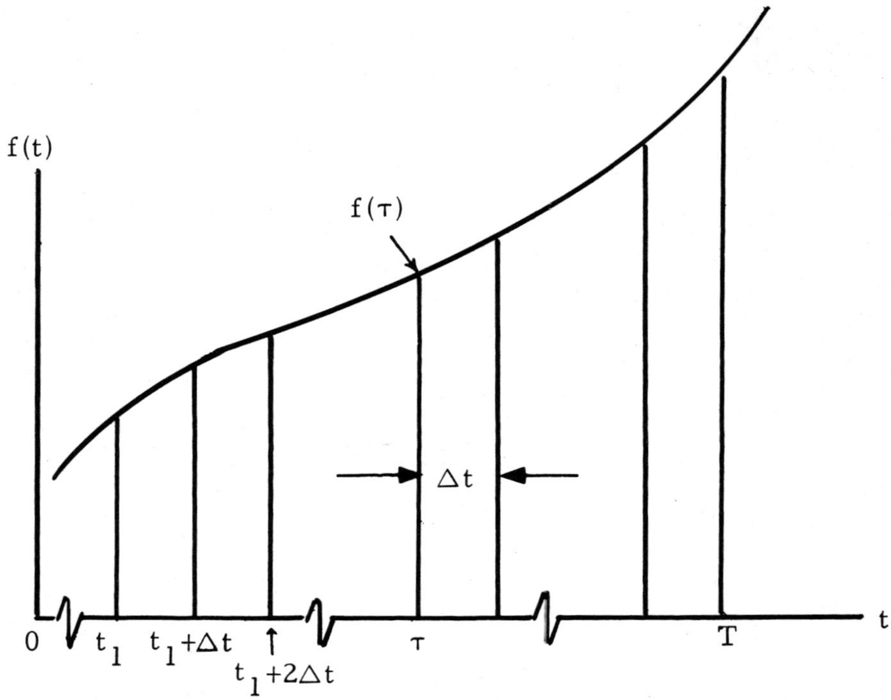

Figure 6-13. The Function $f(t)$ as a Series of Impulse Functions

$h(t)$ = response of the system as a function of time t to unit impulse function $\delta(o)$ applied at time $t = o$

$\delta(t)$ = unit impulse function applied at time t.[a]

From Figure 6-13, we can write the response of the system to an impulse function applied at t_1 and continuing from there on to time t as[b]

$$f(t_1)\,\Delta t h(t - t_1) \tag{6.56}$$

[a] The impulse function $\delta(t)$ is defined by the relations

$$\delta(t - a) = 0 \quad t \neq a$$

$$\int_{a-\epsilon}^{a+\epsilon} \delta(t - a)\,dt = 1 \; \epsilon > 0$$

The first equation indicates that $\delta(t - a)$ is zero except at $t = a$. The second equation requires that the impulse function have unit area.

[b] Note that in (6.56), $f(t_1)\,\Delta t$ represents impulse area.

Similarly, the response of the impulse function applied at $(t_1 + \Delta t)$ and continuing to time t becomes

$$f(t_1 + \Delta t)\,\Delta t h[t - (t_1 + \Delta t)] \tag{6.57}$$

Finally, the response of the system to an impulse applied at τ and continuing to time t can be written

$$f(\tau)\,\Delta t h(t - \tau) \tag{6.58}$$

From equation (6.58) the total response of the system, using superposition property of linear systems, can be obtained by the summation:

$$g(t) = \sum_{\tau=t_1}^{t_2} f(\tau) h(t - \tau)\,\Delta t \tag{6.59}$$

which is the summation form of integral equation (6.55).

To write equation (6.59) in the previously given Fourier notation, let us define the following

$$N = K; \quad t_2 = T; \quad \Delta t = T/N; \quad t_1 = 0 \tag{6.60}$$

$$g(t) = g(t_k) = g_k$$

Using this notation, (6.59) becomes

$$g_k = \frac{T}{N} \sum_{n=0}^{N-1} f_n h_{k-n} \tag{6.61}$$

The relation in (6.61) gives the response of the system at discrete time intervals k for the input function f, having been given the impulse response of the system $h(t)$. As mentioned before, the Fourier transform of g_k will result in the frequency response of the system. Equation (6.61) can also be written in the form

$$g_k = \frac{T}{N} \sum_{n=0}^{N-1} f_{k-n} h_n \tag{6.62}$$

where subscripts for f and h are interchanged. Equivalency of equations (6.61) and (6.62) can be shown by expanding these summations and following the line of reasoning used in deriving equation (6.59).

In digital filter design problems, usually, the desired frequency response of the filter is specified[5] as shown in Figure 6-14. This figure specifies a bandpass filter which passes frequencies between Ω_1 and Ω_2. Using this information, and dividing the frequency range into N equal increments of $\Delta\omega$ each, we can evaluate the impulse response of the system $h(t)$ from equation (6.29),

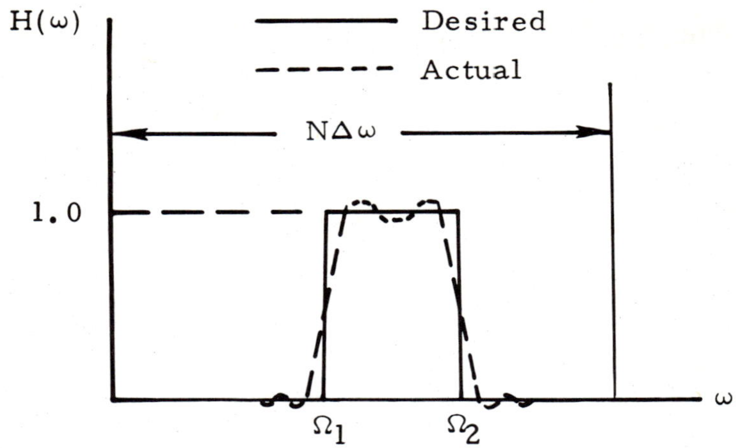

Figure 6-14. Bandpass Filter Passing Frequencies Between Ω_1 and Ω_2

$$h_k = \frac{1}{N} \sum_{n=0}^{N-1} H_n e^{(2\pi j/N)nk} \quad k = 0, \ldots, N - 1 \tag{6.63}$$

where the values of H_n are obtained from Figure 6-14. We may also use the integral form of (6.63), given by (6.20), if $H(\omega)$ can be represented analytically. Having obtained the values of h_k, we can evaluate the time response of the system to any given time function $f(t)$ (in Figure 6-15) by using equations (6.61) or (6.62) in the form,

$$g_k = \frac{T}{N} \sum_{n=o}^{N-1} f_n h_{k-n} = \frac{T}{N} \sum_{n=o}^{N-1} f_{k-n} h_n \tag{6.64}$$

$$k = o, \ldots, N - 1$$

where the time duration T is also divided into N equal increments of Δt each. The desired filter frequency response is shown in Figure 6-14, the digital filter characteristics, however, because of the limited number of steps N will be somewhat different. The mechanized digital filter frequency characteristics can be obtained by calculating the Fourier transform of $h(t)$ given in (6.63). Using equation (6.30) to accomplish this, we get

$$\overline{H(\omega)} = \overline{H}_n = \sum_{k=o}^{N-1} h_k e^{(-2\pi j/N)nk} \tag{6.65}$$

where bars indicate actual filter frequency characteristics accounting for the number of frequency/time steps. Results of similar calculation can be found.[6] Figure 6-14 gives a typical frequency characteristic obtained from evaluation of equation (6.65).

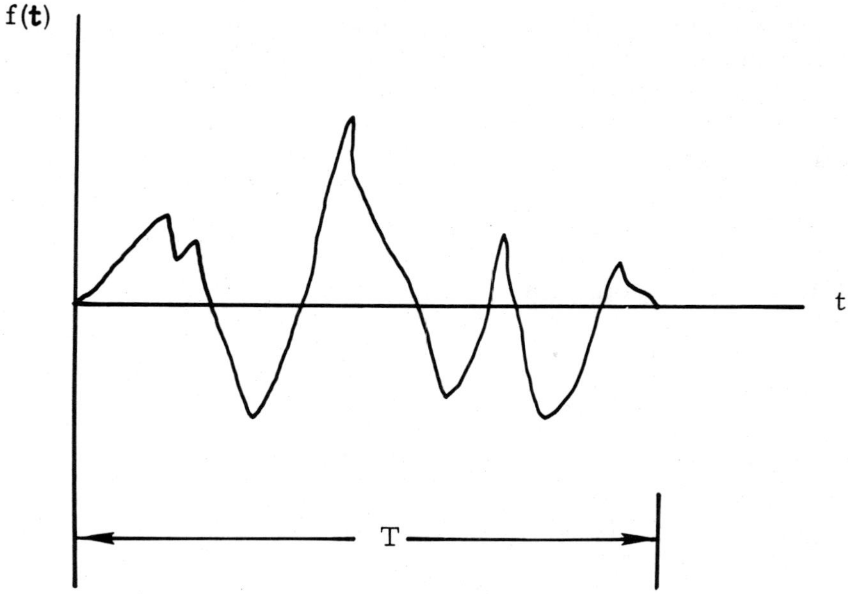

Figure 6-15. Time Function $f(t)$ vs. t

The discussion in this chapter of digital filtering was primarily given to illustrate the use of Fourier transform in filter design. Computational advantages of fast Fourier transform were not included in this discussion. In actual applications, however, evaluation of equations (6.63) and (6.65) are accomplished through the use of fast Fourier transform methods.

In summary, given a typical time function $f(t)$, the discrete digital filter equation (6.64) eliminates the frequency contents of this time function which are outside the bandpass frequency region of $\Omega_1 \leq \omega \leq \Omega_2$ and passes frequencies within this region.

Frequency Consideration

The digital filtering problem can also be considered from a completely different point of view. Assume that the desired filter frequency characteristic is given, as before, by $H(\omega)$ of Figure 6-14. The input signal in frequency domain is also respresented by $F(\omega)$. The purpose of the filter is to eliminate frequencies outside Ω_1 and Ω_2 from the input signal $F(\omega)$. This can be achieved by obtaining the product $F(\omega)H(\omega)$. The desired output of the filter, for input $F(\omega)$, then becomes

$$G(\omega) = F(\omega)H(\omega) \tag{6.66}$$

We will proceed to show that the above equation is the frequency domain representation of the convolution integral (6.55) and, in fact, the previously outlined procedure is merely mechanization of (6.66).

Consider the convolution integral in its general form, with limits of integration between minus and plus infinity,

$$g(t) = \int_{-\infty}^{\infty} f(\tau)h(t - \tau)\, d\tau \tag{6.67}$$

The Fourier transform of (6.67) using equation (6.21) results,

$$G(\omega) = \int_{-\infty}^{\infty}\int_{-\infty}^{\infty} f(\tau)h(t - \tau)e^{-j\omega t}\, d\tau\, dt \tag{6.68}$$

where the constant in front of equation (6.21) is neglected in the above derivation. Replacing $t - \tau$ by x in (6.68) we get

$$G(\omega) = \int_{-\infty}^{\infty} f(\tau)e^{-j\omega\tau}\, d\tau \int_{-\infty}^{\infty} h(x)e^{-j\omega x}\, dx$$

$$= F(\omega)H(\omega) \tag{6.69}$$

where equation (6.21) is again used in obtaining (6.69). Equation (6.69) is the desired result represented by (6.66). Thus, the frequency response of the filter to an input $F(\omega)$ is the product of this input and the frequency response of the filter $H(\omega)$.

Discussion of the Methods

Digital computers have enjoyed an exponential expansion after World War Two and have been applied to the solution of a variety of engineering problems. The digital computers were readily applied to the solution of problems whose formulation were primarily in the *time* domain. These computers were not originally used to solve frequency-domain problems such as tuned filters because the frequency-domain problems had to be converted to time domain and subsequently solved on the computer. This conversion, however, requires high speed analog/digital converters followed by an enormous amount of calculations.

For example, assume it is required to obtain the frequency contents of a waveform containing frequencies in the order of 100 kHz. The sampling rate should at least be twice the highest frequency or in the order of 200 kHz. This will require an analog/digital converter with a 200-kHz rate of handling appropriate number of bits (to achieve the desired significant figure contents). Following analog/digital conversion, the sampled values are put through the Fourier transform equations, resulting in the frequency contents of the original waveform.

This procedure in the past has been carried out by analog filters by which the presence of a given frequency is detected in the corresponding tuned filter. Presently, the availability of analog/digital converters, Cooley-Tukey-type computer algorithms, and high-speed digital computers make the digital signal processing economically attractive.

Problems

1. Obtain the Fourier series for the waveform of example 1 given in Figure 6-8.
2. Consider the solution of example 1 given in Table 6-1. Expand the inversion summation of equation (6.26) in the form

$$F(t) = \Delta\omega\{(a_0 + jb_0)[\cos \omega_0 t + j \sin \omega_0 t] + (a_1 + jb_1)[\cos \omega_1 t + j \sin \omega_1 t] + \ldots\}$$

   Using the values of constants from Table 6-1, obtain the Fourier series of $F(t)$.
3. Obtain the values of W^{nk} for the case of $n = 0, 1, \ldots, 7$ and $k = 0, 1, \ldots, 7$.
4. Write the equivalent matrices to (6.36) and (6.37) for $N = K = 8$.
5. Obtain the product of **W** matrices of equation (6.43) and show that this product is equal to the **W** matrix of (6.42).
6. Using Cooley-Tukey devivation, obtain the order of the **G** matrix for $N = K = 16$.
7. In a computation tree for $N = 16$, find the power of W on row 6 and column 3.
8. In a computation tree for $N = 32$, find the power of W on row 13 and column 3.
9. For $N = 16$, determine the origin of solid and dashed lines coming to the node at row 7, column 2.
10. For $N = 32$, determine the origin of the solid and dashed lines coming to the node at row 18 and column 1.
11. Derive the analytical derivation of the Cooley-Tukey method for $N = 4$.
12. Assume the impulse response of a system to be given

t	$h(t)$
0	0
0.1	0.2
0.2	0.4
0.3	0.6
0.4	0.8
0.5	1.0

Using these data and $\Delta t = 0.1$, evaluate the response of the system to forcing function

t	$f(t)$
0	1.0
0.1	1.3
0.2	1.5
0.3	1.8
0.4	2.5
0.5	3.0

Plot the results.

Notes

1. R.V. Churchill, *Fourier Series and Boundary Value Problems* (New York: McGraw-Hill, 1963).
2. S.A. Hovanessian and Louis A. Pipes, *Digital Computer Methods in Engineering* (New York: McGraw-Hill, 1969), chap. 4.
3. J.W. Cooley and J.W. Tukey, "An Algorithm for Machine Calculation of Complex Fourier Series," *Mathematics of Computation* (April 1965).
4. S.A. Hovanessian and D.G. Kruska, "Calculation of Fourier Transform Using Symmetrical Properties," *International Journal of Computers and Electrical Engineering*, vol. 1 (1974), pp. 467-480.
5. S.D. Stearns, *Digital Signal Analysis*, Hayden Book Co., (Rochelle Park, N.J.: 1975).
6. Ibid., p. 108

Bibliography

Brigham, E. Oran. *The Fast Fourier Transform*. Englewood Cliffs, N.J.: Prentice-Hall, 1974.

"Digital Signal Processing," *Proceedings of the IEEE* (April 1975). Special issue.

Hovanessian, S.A.: Pipes, L.A. *Digital Computer Methods in Engineering*, New York, McGraw-Hill, 1969, chapter 4.

Popoulis, A. *The Fourier Integral and Its Applications*. New York: McGraw-Hill, 1962.

Stearns, S.D. *Digital Signal Analysis*. Rochelle Park, New Jersey: Hayden, 1975.

7 Computer Optimization Methods

In calculus we learned how to obtain the minimum and the maximum of a function by setting its derivative equal to zero. Unfortunately minimization or maximization (optimization) problems encountered in industry are not that simple. Usually, optimization should take place while satisfying a number of constraints imposed on the system. In cases where the constraints and the function to be optimized are expressed analytically, the Lagrangian method of undetermined multipliers can be used to obtain the optimum solution.

Linear programming methods developed in the 1950s can be used very effectively in cases where both the constraints and the function to be optimized are linear. Variations of linear programming methods are also available for cases where the function to be optimized is quadratic (quadratic programming) and for cases where the nonlinear constraints can be expressed as piecewise linear functions (separable programming).

Optimization with Constraints

Consider the function

$$F = f(x, y, z) \tag{7.1}$$

to be optimized (maximum or minimum) subject to the constraint

$$G = g(x, y, z) = 0 \tag{7.2}$$

The total derivative of (7.1) is

$$dF = \frac{\partial f}{\partial x}dx + \frac{\partial f}{\partial y}dy + \frac{\partial f}{\partial z}dz = 0 \tag{7.3}$$

where $\partial f/\partial x$, $\partial f/\partial y$, and $\partial f/\partial z$ are partial derivatives of the function $f(x, y, z)$ with respect to x, y, and z.

The total differential of G can also be written

$$dG = \frac{\partial g}{\partial x}dx + \frac{\partial g}{\partial y}dy + \frac{\partial g}{\partial z}dz = 0 \tag{7.4}$$

Multiplying (7.4) by Lagrangian multiplier λ and adding to (7.3)

$$\left(\frac{\partial F}{\partial x} + \lambda\frac{\partial g}{\partial x}\right)dx + \left(\frac{\partial f}{\partial y} + \lambda\frac{\partial g}{\partial y}\right)dy + \left(\frac{\partial f}{\partial z} + \lambda\frac{\partial g.}{\partial z}\right)dz = 0 \tag{7.5}$$

If λ is chosen so that

$$\begin{aligned}\frac{\partial f}{\partial x} + \lambda\frac{\partial g}{\partial x} &= 0\\ \frac{\partial f}{\partial y} + \lambda\frac{\partial g}{\partial y} &= 0\\ \frac{\partial f}{\partial z} + \lambda\frac{\partial g}{\partial z} &= 0\\ g(x, y, z) &= 0\end{aligned} \tag{7.6}$$

then the necessary condition for extremum will be satisfied. Equations (7.6) represent four equations in four unknowns x, y, z, and λ. The solution of these equations will give the values of the variables x, y, z at *maximum* and *minimum* points of the function f.

Example: Find the maximum and the minimum distances from the origin to the curve

$$5x^2 + 6xy + 5y^2 - 8 = 0 \tag{7.7}$$

Solution: The problem can be stated as maximizing (minimizing) the function

$$f(x,y) = x^2 + y^2 \tag{7.8}$$

which represents the square of the distance from the origin to the point x and y, subject to the constraint equation,

$$g(x,y) = 5x^2 + 6xy + 5y^2 - 8 = 0 \tag{7.9}$$

Using Lagrangian formulation of equation (7.6), we get the three equations which include the constraint equation.

$$\begin{aligned}2x + \lambda(10x + 6y) &= 0\\ 2y + \lambda(6x + 10y) &= 0\\ 5x^2 + 6xy + 5y^2 - 8 &= 0\end{aligned} \tag{7.10}$$

Multiplying the first of these equations by y and the second by x and subtracting, we get

$$6\lambda(y^2 - x^2) = 0 \tag{7.11}$$

or $y = \pm x$. Substituting this in the third equation, we get

$$2x^2 = 1, \quad x^2 = 2 \tag{7.12}$$

The first of these gives

$$f = x^2 + y^2 = 1 \qquad \text{(minimum)} \tag{7.13}$$

and the second gives

$$f = 4 \qquad \text{(maximum)} \tag{7.14}$$

which represents the solution of the problem.

Since the Lagrangian method of undetermined multipliers is used in the quadratic programming method to be discussed in this chapter, the following paragraphs will serve as an introduction to the mathematical derivation of Lagrangian functions.

Lagrangian Multipliers Method

To determine the extremal values of the funciton

$$F = f(x_1, x_2, \ldots, x_n) \tag{7.15}$$

subject to constraining conditions

$$g_i(x_1, x_2, \ldots, x_n) = 0 \qquad (i = 1, 2, \ldots, m) \tag{7.16}$$

we form the Lagrangian function

$$L = F + \sum_{i=1}^{m} \lambda_i g_i \tag{7.17}$$

and determine parameters $\lambda_1, \ldots, \lambda_m; x_1, \ldots, x_n$ from the n equations

$$\frac{\partial L}{\partial x_j} = 0 \qquad (j = 1, \ldots, n) \tag{7.18}$$

and m constraint equations (7.16).

Necessary and Sufficient Condition for an Extremum

Let the problem be stated

$$\text{Maximumize } J = f(x_1, x_2, \ldots x_n) \tag{7.19}$$

where f is of class c^2 (first and second derivatives are continuous functions). The existence conditions of the solution follow.

Necessary Condition

$$f'_{x_i}(x_1\, x_2, \ldots, x_n) = 0 \quad i = 1, \ldots, n \tag{7.20}$$

where f'_{x_i} represents the derivative of the function f with respect to x_i

Sufficient Condition

$$\begin{array}{lll} \text{For minimum} & D_i > 0 & i = 1, \ldots, n \\ \text{For maximum} & D_i > 0 & i = 2, 4, 6, \ldots \\ & D_i < 0 & i = 1, 3, 5, \ldots \end{array} \tag{7.21}$$

where

$$D_i = \begin{vmatrix} f''_{x_1x_1} & f''_{x_1x_2} & \cdots & f''_{x_1x_i} \\ f''_{x_2x_1} & f''_{x_2x_2} & \cdots & f''_{x_2x_i} \\ \cdots & & & \\ f''_{x_ix_1} & \cdots & \cdots & f''_{x_ix_i} \end{vmatrix} \tag{7.22}$$

Equation (7.22) represents the determinant formed by the second derivates of the function f with respect to x_i and x_j variables. The subscript i of determinant D_i also represents the order of the determinant. For example,

$$D_1 = f''_{x_1x_1} \tag{7.23}$$

is the first-order determinant formed from the diagonal element of D_i of equation (7.22), and

$$D_2 = \begin{vmatrix} f''_{x_1x_1} & f''_{x_1x_2} \\ f''_{x_2x_1} & f''_{x_2x_2} \end{vmatrix} \tag{7.24}$$

is the second-order determinant formed from (7.22).

For a *two-dimensional* case these conditions are equivalent to the necessary and sufficient conditions that follow.

Necessary Condition

$$\begin{aligned} f'_{x_1}(x_1, x_2) &= 0 \\ f'_{x_2}(x_1, x_2) &= 0 \end{aligned} \tag{7.25}$$

Sufficient Condition

For minimum $D_1 = f''_{x_1x_1} > 0$

$$D_2 = f''_{x_1x_1}f''_{x_2x_2} - (f''_{x_1x_2})^2 > 0 \tag{7.26}$$

For maximum $D_1 = f''_{x_1x_1} < 0$

$$D_2 = f''_{x_1x_1}f''_{x_2x_2} - (f''_{x_1x_2})^2 > 0 \tag{7.27}$$

where D_1 and D_2 represent the values of the determinant (7.22).

In using the Lagrangian method of undetermined multipliers we are, in effect, maximizing or minimizing the Lagrangian function L given in (7.17). Thus the necessary and sufficient conditions will apply to the Lagrangian function L as given in the following example.

Example: Consider the extremal problem

$$f(x, y) = 2x + 3y - 1 \tag{7.28}$$

subject to constraint

$$g(x, y) = x^2 + 1.5y^2 - 6 = 0 \tag{7.29}$$

Form the Lagrangian $L = f + \lambda g$ as

$$L = 2x + 3y - 1 + \lambda(x^2 + 1.5y^2 - 6) \tag{7.30}$$

and obtain

$$\frac{\partial L}{\partial x} = 2 + 2\lambda x = 0$$

$$\frac{\partial L}{\partial y} = 3 + 3\lambda y = 0 \tag{7.31}$$

The preceding equations, together with the constraint $g(x, y) = 0$, can be solved for x, y, and λ. We get the following

$$x = y = 1/\lambda \tag{7.32}$$

and

$$2.5 - 6\lambda^2 = 0; \quad \lambda = \pm 0.645 \tag{7.33}$$

Using the values of $\lambda = \pm 0.645$, we can compute

$$x = 1/\lambda = 1.55 \text{ and } -1.55 \tag{7.34}$$

and similarly for

$$y = 1/\lambda = 1.55 \text{ and } -1.55 \tag{7.35}$$

Table 7-1
Necessary and Sufficient Conditions for an Extremum

λ	−0.645	0.645
x	1.55	−1.55
y	1.55	−1.55
f	6.75	−8.75
$D_1 = L''_{xx}$	−1.29 ≤0	1.29 >0
$D_2 = L''_{xx}L''_{yy} - (L''_{xy})^2$	2.5 >0	2.5 >0
	Maximum	Minimum

The derivatives of the Lagrangian L are as follow:

$$L''_{xx} = 2\lambda \quad L''_{xy} = L''_{yx} = 0, \quad L''_{yy} = 3\lambda \tag{7.36}$$

Examine the values of the Table 7-1 for necessary and sufficient conditions for an extremum.

From Table 7-1, it is concluded that the maximum value of the function $f = 6.75$ occurs at $x = y = 1.55$. The minimum value of the function $f = -8.75$ occurs at $x = y = -1.55$.

If the constraints $g(x, y)$ or the function $f(x,y)$ contain high powers of the variables $(x_1, \ldots, x_n)$, the Lagrangian formulation will result in nonlinear sets of equations, many times impossible to solve.

Linear Programming Methods

The linear programming method obtains solutions to optimization problems where the functional and the constraints are linear functions of the variables. In these problems we usually have more unknowns than equations. The final solution results in the nonzero values of some of the unknowns, with the rest of the unknowns set equal to zero. Consider the following linear programming formulation:

Constraints

$$\sum_{j=1}^{m} a_{ij}x_j = b_i \quad \begin{cases} i = 1, 2, \ldots, n \\ j = 1, \ldots, m \end{cases} \tag{7.37}$$

Functional

$$\sum_{j=1}^{m} a_{ij}x_j = \min \quad i = 0 \tag{7.38}$$

Nonnegative

$$x_j > 0 \qquad j = 1, \ldots, m \tag{7.39}$$

where $m < n$ for $m = 3$ and $N = 2$, expansion of the equations will result in

Constraints

$$\begin{cases} a_{11}x_1 + a_{12}x_2 + a_{13}x_3 = b_1 \\ a_{21}x_1 + a_{22}x_2 + a_{23}x_3 = b_2 \end{cases} \tag{7.40}$$

Functional to be optimized

$$a_{01}x_1 + a_{02}x_2 + a_{03}x_3 \tag{7.41}$$

Nonnegativity condition

$$x_1, x_2, x_3 \geq 0 \tag{7.42}$$

Note also the nonnegativity condition of the variables x_1, x_2, and x_3. Linear programming problems are usually formulated with this condition for convenience and uniformiity of programmed codes. A negative variable can, of course, be represented as the difference of two positive variables.

Simplex Method

The simplex method is the most popular algorithm for determining a solution to a linear programming problem. It makes use of the fact that the optimum solution will always be at a corner of the feasible region formed by the constraints. Such corner solutions are called *basic* solutions. The simplex algorithm tests subsequent basic solutions until no further reduction (or increase) in the objective function is possible.

As an illustration of the simplex method, consider the following problem:

$$\begin{aligned} 2x_1 - x_2 &\leq 8 \\ -x_1 + 3x_2 &\leq 6 \end{aligned} \tag{7.43}$$

$$x_1 + 4x_2 = \text{maximum} \tag{7.44}$$

We put this problem in linear programming formulation by adding *slack variables* x_3 and x_4 and minimizing the *negative* of the functional.

$$\begin{cases} 2x_1 - x_2 + x_3 = 8 \\ -x_1 + 3x_2 + x_4 = 6 \end{cases} \tag{7.45}$$

$$\begin{cases} -x_1 - 4x_2 = F = \text{minimum} \\ x_1, x_2, x_3, x_4 \geq 0 \end{cases} \tag{7.46}$$

(If some variables are expected to assume negative values, this can be accommodated in the linear programming formulation by setting them as the difference of two positive variables. For example, negative z can be written $z = x - y$ where x and y are positive variables.) For the basis (basic solution) assume

$$x_1 = 0, \quad x_2 = 0, \quad x_3 = 8, \quad x_4 = 6 \tag{7.47}$$

with

$$F = 0 \tag{7.48}$$

The functional F can be lowered by increasing the values of x_1 or x_2 especially x_2 which has the larger *negative* coefficient. Increasing x_2 in constraints where x_2 appears with a positive coefficient will cause the other variables to assume lower numbers and possibly go negative. To avoid this, we consider the constraints where x_2 appears with a positive coefficient and calculate the limiting value of x_2. We bring x_2 into the *basis* through the second constraint as follows:

$$\begin{aligned} 2x_1 - x_2 + x_3 &= 8 \\ -1/3x_1 + x_2 + 1/3x_4 &= 2 \\ -x_1 - 4x_2 &= F \end{aligned} \tag{7.49}$$

Eliminating x_2 in the first constraint and the functional, we get

$$\begin{cases} 5/3x_1 + x_3 + 1/3x_4 = 10 \\ -1/3x_1 + x_2 + 1/3x_4 = 2 \\ -7/3x_1 + 4/3x_4 = F + 8 \end{cases} \tag{7.50}$$

New *basis*: $x_1 = 0$, $x_2 = 2$, $x_3 = 10$, $x_4 = 0$, and $F = -8$.
Now bring x_1 into *basis* through constraint 1:

$$\begin{cases} x_1 + 3/5x_3 + 1/5x_4 = 6 \\ -1/3x_1 + x_2 + 1/3x_4 = 2 \\ -7/3x_1 + 4/3x_4 = 8 \end{cases} \tag{7.51}$$

where the coefficient of x_1 in the first equation is set equal to one by dividing the first equation of (7.50) by 5/3.

Eliminating x_1 from second constraint and the functional, we get

$$\left\{\begin{array}{llll} x_1 & + 3/5x_3 & + 1/5x_4 & = \quad 6 \\ x_2 & + 1/5x_3 & + 2/5x_4 & = \quad 4 \\ & 7/5x_3 & + 41/15x_4 & = F + 22 \end{array}\right. \tag{7.52}$$

Since no negative coefficients appear in the functional the solution is

$$\begin{array}{ll} x_1 = 6, & x_2 = 4 \\ x_3 = 0, & x_4 = 0 \end{array} \tag{7.53}$$

with the functional

$$F = -22 \tag{7.54}$$

Simplex logic used in the solution was developed after World War Two for linear programming problems. Simplex logic and its variations have been programmed for digital computers, and these programs are available at almost every computer installation. These programs have practically no limit on the number of constraints or the number of unknowns used in the linear programming formulation.

Graphical Presentation

Figure 7-1 shows the graphic presentation of the solution obtained. The constraint equation, together with the functional F, are shown as straight lines on this figure. The arrows show the direction in which the basic solution progresses until it reaches the optimal solution (i.e., $x_1 = 6$, $x_2 = 4$). It is seen that this solution is at the intersection of the two constraints, meaning that the solution is in the allowable region. The ellipsoid of the quadratic programming solution is given in the figure for the future discussion of quadratic programming.

Simplex Method Logic

The following logic constitutes the basis of simplex algorithm:

1. Find x_i with the highest negative coefficient in functional.
2. Locate constraints where x_1 has positive coefficients.
3. Divide the right side by the x_1 coefficient.
4. Choose the constraint with the smallest value of 3.
5. Divide this constraint by the coefficient of x_1 and elimnminate x_1 from other constraints and the functional.
6. Repeat the procedure until all coefficients of x_1 in the functional are positive.

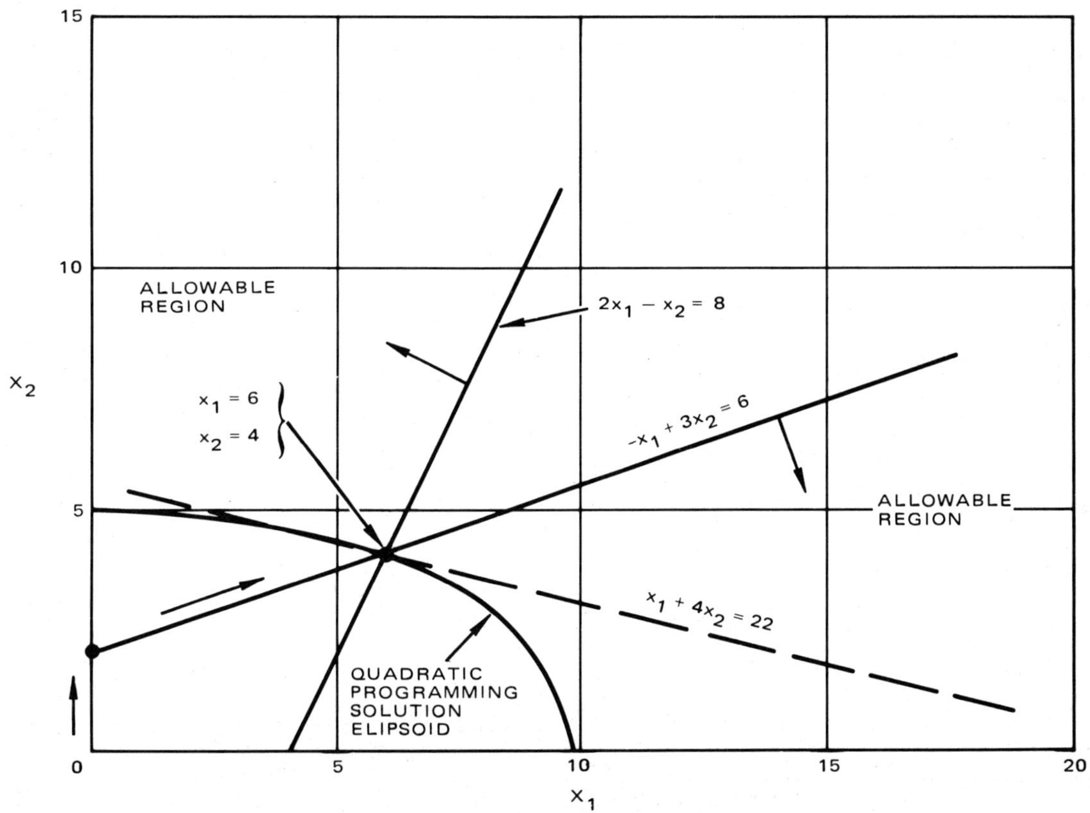

Figure 7-1. Graphical Presentation of Linear Programming Solution

Lagrangian Multipliers and Quadratic Programming

Consider the formulation of the Lagrangian method of undetermined multipliers given previously, but this time altered slightly to put it in the linear programming context. This formulation should result in positive values of Lagrangian multipliers λ as required by available linear programming codes. This Lagrangian formulation follows.

We want to minimize the function

$$F = f(x_1, \ldots, x_n) \tag{7.55}$$

subject to "greater than" constraints

$$g_i(x_1, \ldots, x_n) \geq 0 \qquad i = 1, 2, \ldots, n \tag{7.56}$$

where n is the number of constraints. The Lagrangian function can be formed as discussed previously.

$$L = F - \sum_{i=1}^{k} \lambda_i g_i \tag{7.57}$$

It can be shown that under these conditions the Lagrangian multipliers will assume positive ($\lambda_i \geq 0$) values. Since linear programming formulations result in *nonnegative* variables, we should make sure that the problem is formulated according to the above constraint relations.

Equality constraints can be put in the "greater than" form as follows: consider the equality constraint

$$g(x_1, \ldots, x_n) = 0 \tag{7.58}$$

This constraint can be replaced by two inequality constraints,

$$g(x_1, \ldots, x_n) \geq 0 \quad \text{and} \quad g(x_1, \ldots, x_n) \leq 0 \tag{7.59}$$

The inequality constraints (one "greater than" and one "less than") can be replaced by two "greater than" constraints by multiplying the second one by minus one,

$$\begin{aligned} g(x_1, \ldots, x_n) &\geq 0 \\ -g(x_1, \ldots, x_n) &\geq 0 \end{aligned} \tag{7.60}$$

In the Lagrangian formulation, we should also include the condition that either $g_i = 0$ and $\lambda_i \neq 0$ *or* $g_i \neq 0$ and $\lambda_i = 0$. This condition specifies the fact that one or the other (λ or g) are zero and guarantees that the summation term of the Lagrangian equation (7.57) will not be "undefined." When using a simplex-type computer code to solve a quadratic programming problem, it is usually necessary to add these conditions.

Quadratic Programming in Simplex Formulation

Consider the following optimization problem in variables x_1 and x_2:

$$\text{Minimize } F = a_1x_1 + a_2x_2 + a_{11}x_1^2 + a_{12}x_1x_2 + a_{22}x_2^2 \qquad (7.61)$$

The function F includes *only* first- and second-order terms. The fact that it does not include higher-order terms makes it quadratic. The constraints, however, are linear and include *only* first-order terms:

$$g_1 = b_{11}x_1 + b_{12}x_2 + d_1 \geq 0$$

$$g_2 = b_{21}x_1 + b_{22}x_2 + d_2 \geq 0$$

$$g_3 = x_1 \geq 0$$

$$g_4 = x_2 \geq 0 \qquad (7.62)$$

where the last two equations represent the nonnegativity condition on variables x_1 and x_2. Using equation (7.57), we form the Lagrangian function.

$$\begin{aligned} L &= F - \sum \lambda_i g_i \\ &= a_1x_1 + a_2x_2 + a_{11}x_1^2 + a_{12}x_1x_2 + a_{22}x_2^2 \\ &\quad -\lambda_1 g_1 - \lambda_2 g_2 - \lambda_3 g_3 - \lambda_4 g_4 \end{aligned} \qquad (7.63)$$

We also obtain the following partial derivatives:

$$\frac{\partial L}{\partial x_1} = a_1 + 2a_{11}x_1 + a_{12}x_2 - \lambda_1(b_{11}) - \lambda_2(b_{21}) - \lambda_3(1) = 0$$

$$\frac{\partial L}{\partial x_2} = a_2 + a_{12}x_1 + 2a_{22}x_2 - \lambda_1(b_{12}) - \lambda_2(b_{22}) - \lambda_4 = 0$$

$$\frac{\partial L}{\partial \lambda_1} = -g_1 = 0$$

$$\frac{\partial L}{\partial \lambda_2} = -g_2 = 0$$

$$\frac{\partial L}{\partial \lambda_3} = -g_3 = 0$$

$$\frac{\partial L}{\partial \lambda_4} = -g_4 = 0 \qquad (7.64)$$

Now we form the optimization function by adding the functional F to both sides of the equation (7.63) in the form

$$L + F = 2a_1x_1 + 2a_2x_2 + 2a_{11}x_1^2 + 2a_{12}x_1x_2 + 2a_{22}x_2^2 - \lambda_1 g_1 - \lambda_2 g_2 - \lambda_3 g_3 - \lambda_4 g_4 \tag{7.65}$$

As will be seen this will not alter the character of the optimization problem. It will, however, help us to reduce the quadratic problem to the linear problem.

Solving the first two of equations (7.64) for the values of λ_3 and λ_4 and substituting in equation (7.65), we obtain

$$L + F = a_1x_1 + a_2x_2 - \lambda_1 d_1 - \lambda_2 d_2 \tag{7.66}$$

Note the elimination of the quadratic terms x_1^2, x_1x_2, x_2^2 in going from equation (7.65) to (7.66) with the λ_3 and λ_4 substitution.

The set of equations (7.64) together with the minimization of $L + F$ as given by equation (7.66) constitute the linear programming formulation of the quadratic programming problem. The last two equations of the set (7.64) relate the nonnegativity conditions on x_1 and x_2. These two equations can be dropped from formulation if one keeps track of the condition $\lambda_3 \neq 0$ if $g_3 = 0$ or $\lambda_3 = 0$ if $g_3 \neq 0$ and similarly for λ_4.

Numerical Example

Consider the previously given linear programming example, converted into a quadratic programming problem:

$$\text{Constraints} \quad \begin{cases} 2x_1 - x_2 \leq 8 \\ -x_1 + 3x_2 \leq 6 \end{cases} \tag{7.67}$$

$$\text{Maximize} \quad F = x_1^2 + 4x_2^2 \tag{7.68}$$

Note the quadratic form of the functional F. Putting (7.67) and (7.68) in the formulation of equations (7.61) and (7.62), we get

$$\begin{aligned} g_1 &= 8 - 2x_1 + x_2 \geq 0 \\ g_2 &= 6 + x_1 - 3x_2 \geq 0 \end{aligned} \tag{7.69}$$

$$\begin{aligned} g_3 &= x_1 \geq 0 \\ g_4 &= x_2 \geq 0 \end{aligned} \tag{7.70}$$

and the functional to be minimized

$$F = -x_1^2 - 4x_2^2 \tag{7.71}$$

The Lagrangian function becomes

$$L = -x_1^2 - 4x_2^2 - \lambda_1(8 - 2x_1 + x_2) - \lambda_2(6 + x_1 - 3x_2) - \lambda_3 x_1 - \lambda_4 x_2 \tag{7.72}$$

We also form the following partial derivatives:

$$\frac{\partial L}{\partial x_1} = -2x_1 + 2\lambda_1 - \lambda_2 - \lambda_3 = 0$$

$$\frac{\partial L}{\partial x_2} = -8x_2 - \lambda_1 + 3\lambda_2 - \lambda_4 = 0$$

$$\frac{\partial L}{\partial \lambda_1} = -8 + 2x_1 - x_2 = 0$$

$$\frac{\partial L}{\partial \lambda_2} = -6 - x_1 + 3x_2 = 0$$

$$\frac{\partial L}{\partial \lambda_3} = -x_1 = 0$$

$$\frac{\partial L}{\partial \lambda_4} = -x_2 = 0 \tag{7.73}$$

The optimization function corresponding to (7.66) becomes

$$L + F = -8\lambda_1 - 6\lambda_2 \tag{7.74}$$

The addition of the function F to the Lagrangian L in the equations renders it a linear form and does *not* affect the minimization of the functional. This is because the Lagrangian L was to be minimized anyway.

The linear programming formulation of equations (7.73) and (7.74), with the last two of the equations of set (7.73) eliminated, is given in Table 7-2, where variables u_1, u_2, u_3, and u_4 are added to form the initial basis. The basic solution from the tableau of Table 7-2 is given as

$$x_1 = x_2 = \lambda_1 = \lambda_2 = \lambda_3 = \lambda_4 = 0 \tag{7.75}$$

and

$$u_1 = u_2 = 0, \qquad u_3 = 8, \qquad u_4 = 6 \tag{7.76}$$

with $F + L = 0$.

Table 7-2
Linear Programming Formulation

Column	*1*	*2*	*3*	*4*	*5*	*6*	*7*	*8*	*9*	*10*	*Right hand side*
Variables	x_1	x_2	λ_1	λ_2	λ_3	λ_4	u_1	u_2	u_3	u_4	*RHS*
(a)	−2		2	−1	−1		1				0
(b)		−8	−1	3		−1		1			0
(c)	2	−1							1		8
(d)	−1	3								1	6
			−8	−6							$F + L$

The following procedure, although it does not follow the exact pattern of the simplex method, will lead to a solution. Bring λ_2 into basis through (*b*) of the tableau. This will result in the following basis given, at the end of the first iteration.

$$\begin{aligned} \lambda_2 &= 0 \\ u_1 &= 0 \\ u_3 &= 8 \\ u_4 &= 6 \end{aligned} \qquad F + L = 0 \tag{7.77}$$

with the rest equaling zero. Similarly, the basis at the end of the second iteration becomes

$$\begin{aligned} \lambda_1 &= 0 \\ \lambda_2 &= 0 \\ u_3 &= 8 \\ u_4 &= 10 \end{aligned} \qquad F + L = 0 \tag{7.78}$$

with the rest of the variable equaling zero. The third iteration will result in

$$\begin{aligned} x_2 &= 2 \\ \lambda_1 &= 3.2 \\ \lambda_2 &= 6.4 \\ u_3 &= 10 \end{aligned} \qquad F + L = -4 \tag{7.79}$$

And, similarly, the fourth iteration is

$$x_1 = 6$$
$$x_2 = 4 \qquad F + L = -200 \qquad (7.80)$$
$$\lambda_1 = 13.6$$
$$\lambda_2 = 15.2$$

The solution satisfies all of the constraints and results in the values of λ_3 and λ_4 as zero, making it possible for x_1 and x_2 assume nonzero values (see the last two of the set of equation [7.73]). With these values, the optimization function

$$F = -x_1^2 - 4x_2^2 \qquad (7.81)$$

becomes -100. Graphically, the ellipsoid of the quadratic programming solution is shown in Figure 7-1.

Nonlinear Constraints

In case of *nonlinear* constraints the method of separable programming which consists of piecewise linearization can be applied.

Consider the following mathematical programming problem with nonlinear constraints:

$$x_2^2 - x_1 = 0 \qquad (7.82)$$

$$x_1 + 2x_2 = 8 \qquad (7.83)$$

$$x_2 = \text{maximum} \qquad (7.84)$$

A graph of the constraints and the possible solution is shown in Figure 7-2. The *nonlinear* constraint can be piecewise linearized as follows:

$$x_1 = \lambda_1(0) + \lambda_2(1) + \lambda_3(4) + \lambda_4(9) \qquad (7.85)$$

$$x_2 = \lambda_1(0) + \lambda_2(1) + \lambda_3(2) + \lambda_4(3) \qquad (7.86)$$

$$1 = \lambda_1 + \lambda_2 + \lambda_3 + \lambda_4 \qquad (7.87)$$

These equations, together with the condition that no more than two of λs can be nonnegative and these should be *consecutive*, are sufficient to keep the solution on the straight-line approximations. The equation and the stated condition defines the line segments between points (0,0), (1,1), (4,2), and (9,3) of Figure 7-2.

From Figure 7-2, it is noted that, since we have two constraints and two unknowns, the solution must lie on the intersection of these constraints. This graphical solution will also be obtained from the following linear programming formulation.

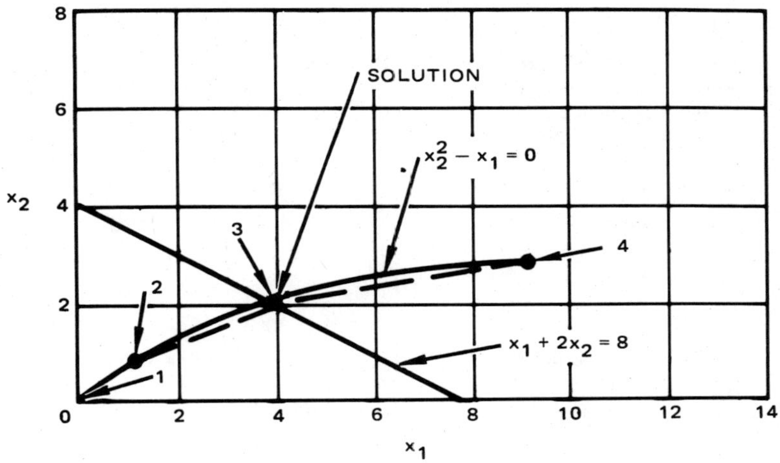

Figure 7-2. Constraints and the Solution of the Example

Table 7-3
Linear Programming Formulation

x_1	x_2	λ_1	λ_2	λ_3	λ_4	*RHS*
−1			1	4	9	0
	−1		1	2	3	0
		1	1	1	1	
1	2					8
	−1					*F* minimize

The linear programming formulation is shown in Table 7-3. The *consecutive* condition on λs will be satisfied automatically, since the x_2 maximization will require the solution to be on the line segments.

This means that the solution will not *jump* between points. For example, the line segment between points **1** and **4** will result in a lower value of the functional, x_2, than the line segments between **1** - **2**, **2** - **3** and **3** - **4**.

The solution of the tableau in Table 7-3 can be accomplished by the simplex method. This was done by using an available computer program. The results of the computer run are given on Table 7-4. Note that this computer program is written to maximize the functional.

The solution at the end of the computer output is as follows,

Table 7-4
Solution of Separable Programming Problem

Your Variables 1 through 6
Artificial Variables 7 through 10

Basis before Iteration 1

Variable	*Value*
7	0
8	0
9	1
10	8

Objective Function Value = 0

Basis before Iteration 2

Variable	*Value*
7	0
6	0
9	1
10	8

Objective Function Value = 0

Basis Before Iteration 3

Variable	*Value*
2	0
6	0
9	1
10	8

Objective Function Value = 0

Basis Before Iteration 4

Variable	*Value*
2	0
4	0
9	1
10	8

Objective Function Value = 0

Basis before Iteration 5

Variable	*Value*
2	1
4	1
1	1
10	5

Objective Function Value = 4

Basis before Iteration 6

Variable	*Value*
2	1.83333
4	0.583333
1	4.33333
6	0.416667

Objective Function Value = 1.8333

Answers

Variable	*Value*
2	2
5	1
1	4
6	0

Objective Function Value = 2

$$
\begin{aligned}
x_1 &= 4 \\
x_2 &= 2 \\
\lambda_3 &= 1 \\
\lambda_4 &= 0 \quad (\lambda_1 = \lambda_2 = 0) \\
\text{Functional} &= 2
\end{aligned}
\tag{7.88}
$$

which is the correct solution given on Figure 7-2.

A variation of the problem where the solution will occur between line segments is obtained by replacing the constraint (7.83) with

$$x_1 + 2x_2 = 4 \tag{7.89}$$

The constraint given by equation (7.89) results in a solution

$$\begin{aligned} x_1 &= 1.6 \\ x_2 &= 1.2 \\ \lambda_2 &= 0.8 \\ \lambda_3 &= 0.2 \quad (\lambda_1 = \lambda_4 = 0) \\ \text{Functional} &= 1.2 \end{aligned} \tag{7.90}$$

This solution *is* on the line segment **2** - **3** of Figure 7-2.

Problems

1. Obtain the necessary and sufficient conditions for the optimization problem with the constraint

$$g(x,y) = x^2 + 2y^2 + 3xy = 0$$

subject to optimization of

$$f(x,y) = 3xy + y$$

2. Obtain the necessary and sufficient conditions for the optimization problem with the constraint

$$g(x,y) = 2x + y + 3z^2$$

subject to optimization of

$$f(x,y) = x^2 + y^2 + z^2$$

3. Formulate the solution of the optimization problem

$$g(x,y) = 3x^2 + y = 0$$

$$h(x,y) = 3x + y^2 = 0$$

subject to optimization

$$f(x,y) = 4x^2 + 7y^2 + 3xy$$

by the method of Lagrangian multipliers.

4. Obtain the solution of linear programming problem

$$2x_1 - x_2 \leq 8$$

$$-x_1 + 3x_2 \leq 6$$

$$x_1 + 4x_2 = \text{maximum}$$

5. Obtain the solution of the linear programming problem

$$10x_1 - 3x_2 \leq 30$$

$$-x_1 + 2x_2 \leq 4$$

$$x_1 + x_2 \geq 4$$

$$x_1 + x_2 = \text{maximum}$$

6. Obtain the solution of the linear programming problem

$$x_1 - x_2 + 2x_3 - x_4 = 2$$

$$2x_1 + x_2 - 3x_3 + x_4 = 6$$

$$x_1 + x_2 + x_3 + x_4 = 7$$

$$-2x_1 - x_2 + x_3 + x_4 = \text{maximum}$$

7. Obtain the solution of the linear programming problem

$$2x_1 + x_2 \geq 2$$

$$x_1 + 3x_2 \leq 3$$

$$2x_2 \leq 4$$

$$3x_1 - x_2 = \text{maximum}$$

8. Formulate the quadratic optimization problem subject to the constraints

$$2x_1 + x_2 \geq 3$$

$$x_1 + x_2 \leq 16$$

and the minimization function

$$F = x_1^2 + 2x_1x_2$$

9. Formulate the quadratic optimization problem subject to the constraints

$$3x_1 + x_2 \leq 16$$

$$x_1 + x_2 \geq 4$$

and the minimization function

$$F = x_1^2 + 3x_2^2 + x_1x_2$$

10. Consider the nonlinear constraint represented by data points

x	y
1	1
2	4
4	6
6	7
9	6
10	5

Represent this function as a piecewise linear function for separable programming formulation.

11. Assuming surfaces can be represented by a series of triangular planes having equations of the form

$$Ax + By + Cz = 1$$

obtain the values of A, B, and C for a plane with vertices (1,1,1), (2,2,3), and (1,2,2).

12. Give separable programming equations that represent the surface of problem 11.
13. Give the equations of separable programming representing the constraint consisting of a triangular surface

$$Ax + By + Cz = 1$$

with vertices (1,1,1), (2,2,1), and (3,2,1).

14. Carry out the computer solution of the separable programming problem given in the text. Represent complete tableaus for each iteration.

Bibliography

Converse, A.O. *Optimization*. New York: Holt, Rinehart and Winston, 1970. Quadratic programming.

Greenberg, H. *Integer Programming*. New York: Academic Press, 1971.

Hovanessian, S.A., and Pipes, L.A., *Digital Computer Methods in Engineering*. New York: McGraw-Hill, 1969. Linear and dynamic programming.

Kuester, J.L., and Maze, J.H., *Optimization Techniques with Fortran*. New York: McGraw-Hill, 1973.

Pun, L. *Optimization Practice*. New York: Wiley, 1969. General reference.

Rose, D.J., and Willoughby, R.A. *Sparse Matrices*. Plenum Press, 1972. Computational aspects of linear programming.

Sivazlian, B.D., and Stanfel, L.E. *Optimization Techniques in Operation Research*. Englewood Cliffs, N.J.: Prentice-Hall, 1975.

Sokolnikoff, I.S. *Advanced Calculus*. New York: McGraw-Hill, 1939. Or any other advanced calculus book for Lagrangian multipliers.

8 Calculation of Characteristic Values and Vectors

Characteristic values and vectors perform an important function in the solution of vibration problems, as they determine the stability of structures such as buildings, bridges, and dams. Vibration problems are formulated, using Newton's laws of motion, and, depending on the number of degrees of freedom n, result in a characteristic equation of order n. The solution of this characteristic equation results in the modes of vibration together with the natural frequencies corresponding to these modes.

In addition to vibration problems, characteristic values and vectors are used in the solution of matrix differential equations and also in the determination of the stability of solution of certain numerical methods used for the solution of simultaneous linear equations.

In a number of engineering problems, it is desirable to reduce a function involving cross products into its quadratic form through a coordinate transformation. This coordinate transformation, which eliminates cross product terms, is obtained by calculating the characteristic values and vectors of the original function.

General Background

Consider the matrix relationship describing the diagonalization of matrix **A** with elements a_{ij},

$$\mathbf{S}^{-1}\mathbf{A}\mathbf{S} = \mathbf{\Lambda} \tag{8.1}$$

where $\mathbf{\Lambda}$ is a diagonal matrix of order n, and **S** is a square matrix of order n with elements s_{ij}. Matrix $\mathbf{S}^{-1}$ is the inverse of matrix **S**.

Equation (8.1) relates matrix **A**, to the diagonal matrix $\mathbf{\Lambda}$. Matrix **S**, which is needed in the diagonalization process, is called the modal matrix. Determination of the elements of matrix **S** and the elements of the diagonal matrix $\mathbf{\Lambda}$ forms an important part of matrix computations. In fact, most of the discussion in this chapter deals with numerical evaluation of the elements of diagonal matrix $\mathbf{\Lambda}$—characteristic values—and the elements of matrix **S**—characteristic vectors.

Premultiplying equation (8.1) by **S** we get

$$\mathbf{A}\mathbf{S} = \mathbf{S}\mathbf{\Lambda} \tag{8.2}$$

Define $\mathbf{S}_k$ as a column vector having the same components as the kth column of matrix $\mathbf{S}$, i.e.,

$$\mathbf{S}_k = \begin{pmatrix} s_{1k} \\ s_{2k} \\ \vdots \\ s_{nk} \end{pmatrix} = (s_{1k}, s_{2k}, \ldots, s_{nk})' \tag{8.3}$$

where prime denotes transpose matrix. Using this relation, equation (8.2) can be written.

$$\mathbf{AS}_k = \mathbf{S}_k\lambda_k \tag{8.4}$$

or

$$(\mathbf{A} - \lambda\mathbf{I})\mathbf{S}_k = 0 \tag{8.5}$$

where $\mathbf{I}$ is the unity matrix. The equation will be satisfied under the condition that the determinant of the parentheses equals zero,

$$|\mathbf{A} - \lambda\mathbf{I}| = 0 \tag{8.6}$$

For an nth order matrix $\mathbf{A}$, equation (8.6) will result in an nth degree polynomial in λ. The values of λ obtained from this polynomial are called the characteristic values. Having obtained the characteristic values, equation (8.5) will, for each λ, result in a set of homogeneous equations,

$$\begin{aligned} (a_{11} - \lambda)s_{1k} + a_{12}s_{2k} + \ldots &= 0 \\ a_{21}s_{1k} + (a_{22} - \lambda)s_{2k} + \ldots &= 0 \\ a_{n1}s_{1k} + a_{n2}s_{2k} + \ldots &= 0 \end{aligned} \tag{8.7}$$

to be solved for the elements of the characteristic vector $\mathbf{S}_k$, i.e., s_{1k}, s_{2k}, $\ldots$, s_{nk}. Note that, since equations (8.7) are homogeneous, one solution will be when $s_{1k} = s_{2k} = \ldots = s_{nk} = 0$. In addition to this trivial solution, we can solve for the ratios of the elements of the $\mathbf{S}_k$ vector; e.g., s_{1k}/s_{nk}, s_{2k}/s_{nk}, $\ldots$, where each equation of (8.7) is divided by the element s_{nk}. Taking the first $(n - 1)$ equations will result in $(n - 1)$ simultaneous equations in $(n - 1)$ unknowns s_{1k}/s_{nk}, s_{2k}/s_{nk}, $\ldots$, $s_{n-1,k}/s_{nk}$. This normalization can also be accomplished by setting $s_{nk} = 1$ and dropping the last equation. This, in effect, normalizes the components of vector $\mathbf{S}_k$ with respect to the last elements s_{nk}.

Theorem. If matrix $\mathbf{A}$ is real and symmetric, then the roots of the characteristic equation $|\mathbf{A} - \lambda\mathbf{I}| = 0$ are all real.

Similarity transformation is another important matrix transformation which is employed in numerical matrix evaluation. The similarity transformation of matrix **A** is defined as

$$\mathbf{D} = \mathbf{P}^{-1}\mathbf{AP} \tag{8.8}$$

where **P** is a nonsingular matrix. Note the difference between the similarity transformation and the diagonalization of the matrix of equation (8.1). In that equation, the product of matrices $\mathbf{S}^{-1}\mathbf{AS}$ resulted in a diagonal matrix Λ, while in similarity transformation **D** is not necessarily diagonal.

Theorem. The characteristic roots of similar matrices are equal: The proof is as follows:

$$\begin{aligned} |\mathbf{P}^{-1}\mathbf{AP} - \lambda\mathbf{I}| &= |\mathbf{P}^{-1}(\mathbf{A} - \lambda\mathbf{I})\mathbf{P}| = 0 \\ &= |\mathbf{P}^{-1}||(\mathbf{A} - \lambda\mathbf{I}||\mathbf{P}| = 0 \end{aligned} \tag{8.9}$$

or

$$|\mathbf{A} - \lambda\mathbf{I}| = 0 \tag{8.10}$$

This theorem is used in a number of numerical methods for obtaining characteristic values and vectors.

Theorem. Let **A** be a matrix with characteristic values $\lambda_1, \lambda_2, \ldots, \lambda_n$ and the function $F(x) = a_0 + a_1x + a_2x^2 + \ldots + a_mx^m$ to be given. The characteristic values of $F(\mathbf{A})$, where matrix **A** replaces the variable x, will be $F(\lambda_1), F(\lambda_2), \ldots, F(\lambda_n)$.

As a consequence of this theorem, the characteristic values of $\mathbf{A}^k$ will equal λ^k, where λs are the characteristic values of **A**.

Theorem. Let **A** be a real matrix with characteristic values $\lambda_1, \ldots, \lambda_n$. The characteristic values of the *transposed* matrix $\mathbf{A}'$ will also be $\lambda_1, \ldots, \lambda_n$.

This theorem will be used in the derivations of numerical methods given in the following pages.

First Characteristic Value and Vector[a] (POWER METHOD)

Given matrix **A** with characteristic values $\lambda_1, \ldots, \lambda_n$ and characteristic vectors $\mathbf{X}_1, \ldots, \mathbf{X}_n$ where

---

[a] The designations of characteristic value and eigenvalue are used interchangeably throughout this chapter.

$$\mathbf{X}_i = (x_{1i}, x_{2i}, x_{3i}, \ldots, x_{ni})' \tag{8.11}$$

Consider the vector formed from addition of the characteristic vectors of matrix **A**, as

$$\mathbf{Y}_o = a_1\mathbf{X}_1 + a_2\mathbf{X}_2 + \ldots + a_n\mathbf{X}_n \tag{8.12}$$

We may form

$$\mathbf{AY}_o, \mathbf{A}^2\mathbf{Y}_o, \ldots, \mathbf{A}^k\mathbf{Y}_o \tag{8.13}$$

From the diagonalization equation, we have

$$\mathbf{AX} = \Lambda\mathbf{X} \tag{8.14}$$

Replacing **X** in the above equation by $\mathbf{Y}_o$ from (8.12) and noting that Λ represents a diagonal matrix with elements $\lambda_1, \lambda_2, \ldots, \lambda_n$, we have

$$\begin{aligned} \mathbf{AY}_o &= a_1\lambda_1\mathbf{X}_1 + a_2\lambda_2\mathbf{X}_2 + \ldots + a_n\lambda_n\mathbf{X}_n \\ \ldots & \\ \mathbf{A}^k\mathbf{Y}_o &= a_1\lambda_1^k\mathbf{X}_1 + a_2\lambda_2^k\mathbf{X}_2 + \ldots + a_n\lambda_n^k\mathbf{X}_n \end{aligned} \tag{8.15}$$

Denoting the vector on the left side of the last equation by

$$\mathbf{A}^k\mathbf{Y}_o = \mathbf{Y}_k = (y_{1k}, y_{2k}, \ldots, y_{nk})' \tag{8.16}$$

or each component of $\mathbf{Y}_k$ can be equated to respective component of $\mathbf{X}_k$ through equation (8.15), as

$$y_{ik} = a_1x_{i1}\lambda_1^k + a_2x_{i2}\lambda_2^k + \ldots + a_nx_{in}\lambda_n^k \tag{8.17}$$

Thus any component of $\mathbf{Y}_k$ will depend linearly on λ^k or

$$y_k = C_1\lambda_1^k + C_2\lambda_2^k + \ldots + C_n\lambda_n^k \tag{8.18}$$

Leaving this for the moment, consider the following cases of relative characteristic values:

$$|\lambda_1| > |\lambda_2| \tag{8.19}$$

where λ_1 is greater than λ_2 in absolute value, and

$$\lambda_1 = \lambda_2 = \ldots = \lambda_r, \quad |\lambda_r| > |\lambda_{r+1}| \tag{8.20}$$

where λ_1 through λ_r are equal, but λ_r is greater than λ_{r+1}, and

$$\lambda_1 = -\lambda_2, \quad |\lambda_1| > |\lambda_3| \tag{8.21}$$

where $\lambda_1 = -\lambda_2$, but λ_1 is greater than λ_3. In all of these cases, the following relations between components of **Y** obtained at iteration k, (y_k) and at iteration $k + 1$, (y_{k+1}) apply:

$$\frac{y_{k+1}}{y_k} = \frac{C_1\lambda_1^{k+1} + C_2\lambda_2^{k+1} \ldots + C_n\lambda_n^{k+1}}{C_1\lambda_1^k + C_2\lambda_2^k + \ldots + C_n\lambda_n^k}$$

$$= \lambda_1 \frac{1 + b_2\alpha_2^{k+1} + \ldots}{1 + b_2\alpha_2^k + \ldots} \qquad (8.22)$$

where $b_i = C_i/C_1$, $\alpha_i = \lambda_i/\lambda_1$. After dividing through and denoting "the order of" by O, we get

$$\lambda_1 = \frac{y_{k+1}}{y_k} + O\left(\frac{\lambda_2}{\lambda_1}\right)^k \qquad (8.23)$$

Thus, the characteristic value of λ_1 is approximately equal to the ratio of the component of $\mathbf{Y}$ at iteration k and $k + 1$, provided the characteristic value $\lambda_1 > \lambda_2$, so that $(\lambda_2/\lambda_1)^k$ tends to zero.

In each step of the iteration, we may divide the components of the vector $\mathbf{Y}_k$ by the largest element of $\mathbf{Y}_k$ to avoid growth of errors. We obtain λ by the ratio of components $\mathbf{AY}_{k+1}$ and $\mathbf{AY}_k$.

The *characteristic vectors* are given from (8.15) as

$$\mathbf{Y}_k = \mathbf{A}^k\mathbf{Y}_o = \lambda_1^k\left[a_1\mathbf{X}_1 + a_2\left(\frac{\lambda_2}{\lambda_1}\right)^k\mathbf{X}_2 + \ldots + a_n\left(\frac{\lambda_n}{\lambda_1}\right)^k\mathbf{X}_n\right]$$

$$= a_1\lambda_1^k\left[\mathbf{X}_1 + 0\left(\frac{\lambda_2}{\lambda_1}\right)^k\mathbf{X}_2\right] \qquad (8.24)$$

From (8.24), the components of $\mathbf{Y}_k$ are proportional to characteristic vector $\mathbf{X}_1$, provided $(\lambda_2/\lambda_1)^k \ldots (\lambda_n/\lambda_1)^k$ tends to zero with large values of k.

Improving the Convergence of the Iteration Process

Consider the following methods for improving the convergence of the iteration process.

1. *Powers of the matrix*. In this method, we use the previously given theorem, which states that the characteristic values of $\mathbf{A}^m$ will equal λ^m, where λ is the characteristic value of $\mathbf{A}$. Furthermore, we have the relation that the trace of matrix $\mathbf{A}$ (sum of its diagonal elements) is equal to the sum of eigenvalues of $\mathbf{A}$. These theorems for $\mathbf{A}^m$ result in

$$\mathrm{tr}\mathbf{A}^m = \lambda_1^m + \lambda_2^m + \ldots + \lambda_n^m \qquad (8.25)$$

where tr is the trace. Taking the mth root of both sides

$$\sqrt[m]{\text{tr}\mathbf{A}^m} = \lambda_1 \sqrt[m]{1 + \left(\frac{\lambda_2}{\lambda_1}\right)^m + \ldots + \left(\frac{\lambda_n}{\lambda_1}\right)^m}$$

$$= \lambda_1 + O\left(\frac{1}{m}\left(\frac{\lambda_2}{\lambda_1}\right)^m\right) \tag{8.26}$$

Thus, by raising the matrix $\mathbf{A}$ to the mth power, we will be improving the convergence of its characteristic value λ_1 by a factor of m. Note that raising matrices to a power will take n^3 multiplications where n is the order of the matrix.

For example, starting with $\mathbf{A}^8$ and $\mathbf{A}^8\mathbf{Y}_0$, the next iteration will give us $\mathbf{A}^8(\mathbf{A}^8\mathbf{Y}_0) = \mathbf{A}^{16}\mathbf{Y}_0$. We may calculate $\mathbf{A}(\mathbf{A}^{16}\mathbf{Y}_0) = \mathbf{A}^{17}\mathbf{Y}_0$ and λ_1 is found by the ratio of the components of $\mathbf{A}^{17}\mathbf{Y}_0$ and $\mathbf{A}^{16}\mathbf{Y}_0$.

2. *Scalar product method.* Consider matrix $\mathbf{A}$ with the initial characteristic vector $\mathbf{Y}_0$. The iteration method will proceed as follows:

$$\mathbf{Y}_0,\ \mathbf{Y}_1 = \mathbf{A}\mathbf{Y}_0, \quad \mathbf{Y}_2 = \mathbf{A}^2\mathbf{Y}_0, \ldots, \quad \mathbf{Y}_k = \mathbf{A}^k\mathbf{Y}_0 \tag{8.27}$$

In this method, we also compute the characteristic vector of the transposed matrix $\mathbf{A}'$, starting again with $\mathbf{Y}_0$,

$$\mathbf{Y}_0,\ \mathbf{Y}'_1 = \mathbf{A}'\mathbf{Y}_0, \ldots, \quad \mathbf{Y}'_k = \mathbf{A}'^k\mathbf{Y}_0 \tag{8.28}$$

Using the previous reasoning and also the fact that the characteristic values of the transposed matrices are equal, we can write the dot product of $\mathbf{Y}'_k$ and $\mathbf{Y}_k$ as

$$(\mathbf{Y}'_k \cdot \mathbf{Y}_k) = a_1 b_1 \lambda_1^{2k} + \ldots + a_n b_n \lambda_n^{2k} \tag{8.29}$$

and the dot product of $\mathbf{Y}'_{k-1}$ and $\mathbf{Y}_k$ can also be written

$$(\mathbf{Y}'_{k-1} \cdot \mathbf{Y}_k) = a_1 b_1 \lambda_1^{2k-1} + \ldots + a_n b_n \lambda_n^{2k-1} \tag{8.30}$$

The ratio of the dot products will give the desired value of the characteristic value λ_1

$$\frac{(\mathbf{Y}'_k \cdot \mathbf{Y}_k)}{(\mathbf{Y}'_{k-1} \cdot \mathbf{Y}_k)} = \lambda_1 + O\left(\frac{\lambda_2}{\lambda_1}\right)^{2k} \tag{8.31}$$

Note that the convergence of this method is proportional to $2k$ instead of k.

Thus, formation of scalar products cuts the iteration by almost one-half. In the case of symmetrical matrices since $\mathbf{A}' = \mathbf{A}$, this method is very attractive. Beginning at a certain step in iteration, one forms the scalar product and calculates

$$\lambda_1 \cong \frac{(\mathbf{A}^k\mathbf{Y}_0 \cdot \mathbf{A}^k\mathbf{Y}_0)}{(\mathbf{A}^{k-1}\mathbf{Y}_0 \cdot \mathbf{A}^k\mathbf{Y}_0)} \tag{8.32}$$

In this case, the calculations involving the transpose matrix need not be carried out.

Example: Consider the symmetrical matrix

$$\mathbf{A} = \begin{pmatrix} 2 & -1 & 0 \\ -1 & 2 & -1 \\ 0 & -1 & 1 \end{pmatrix} \tag{8.33}$$

It is desired to obtain its characteristic value and vector by the iteration method. Starting the iteration process with an initial vector of

$$\mathbf{Y}_0 = \begin{pmatrix} 1 \\ -1 \\ 1 \end{pmatrix} \tag{8.34}$$

we get

$$\mathbf{Y}_1 = \mathbf{AY}_0 = 2 \begin{pmatrix} 1.5 \\ -2 \\ 1 \end{pmatrix}; \ \mathbf{Y}_2 = \mathbf{AY}_1 = 3 \begin{pmatrix} 1.6666 \\ -2.1666 \\ 1.0 \end{pmatrix} \tag{8.35}$$

where $\mathbf{Y}$ matrix is normalized with respect to the last element. In computing $\mathbf{Y}_2$, the column vector $(1.5, -2, 1)'$ is for $\mathbf{Y}_1$. Thus, we eliminated the normalization constant 2 from further consideration. Continuing with $\mathbf{Y}_2 = (1.666, -2.1666, 1)'$, we get

$$\mathbf{Y}_3 = 3.1666 \begin{pmatrix} 1.7368 \\ -2.2105 \\ 1.0 \end{pmatrix}, \quad \ldots, \quad \mathbf{Y}_{10} = 3.247 \begin{pmatrix} 1.802 \\ -2.247 \\ 1.0 \end{pmatrix} \tag{8.36}$$

and

$$\mathbf{Y}_{11} = 3.247 \begin{pmatrix} 1.802 \\ -2.247 \\ 1.0 \end{pmatrix} \tag{8.37}$$

There is no change between the 10th and 11th iteration values. Since we used $\mathbf{Y}_{10} = (1.802, -2.242, 1.0)'$ to obtain $\mathbf{Y}_{11}$, the ratio of elements of $\mathbf{Y}_{10}$ and $\mathbf{Y}_{11}$ will be 3.247 which, according to equation (8.23), represents the characteristic value. According to equation (8.24), the characteristic vector corresponding to $\lambda_1 = 3.247$ is $(1.802, -2.247, 1.0)'$.

We can also show from these values that the application of scalar products improves the convergence. Since matrix **A** of the example problem is a symmetrical matrix, applying equation (8.32) on vectors $\mathbf{Y}_2$ and $\mathbf{Y}_3$, we get an improved value of λ_1

$$\begin{aligned}\lambda_1 &= 3.1666\frac{\mathbf{Y}_3 \cdot \mathbf{Y}_3}{\mathbf{Y}_2 \cdot \mathbf{Y}_3} \\ &= 3.166\frac{(1.7368)^2 + (2.2105)^2 + (1)^2}{(1.6666)(1.7368) + \ldots + (1)^2} \qquad (8.38) \\ &= 3.2462\end{aligned}$$

This value is close to the answer of 3.247 obtained previously.

Second Characteristic Value and Vector

Assume for matrix **A** we have obtained λ_1 and the corresponding characteristic vector $\mathbf{X}_1 = (x_1, x_2, \ldots, x_n)'$. In this definition, the second subscript on the components of $\mathbf{X}_1$, i.e., $x_1, x_2, \ldots, x_n$, is dropped for ease of writing. We want to calculate the second characteristic value λ_2 and the corresponding vector $\mathbf{X}_2$. This is accomplished by a series of similarity transformations.

Consider matrix **P** formed from the components of vector $\mathbf{X}_1$:

$$\mathbf{P} = \begin{pmatrix} x_1 & 0 & \ldots & 0 \\ x_2 & 1 & \ldots & 0 \\ \vdots & & & \\ x_n & 0 & & 1 \end{pmatrix} \qquad (8.39)$$

The inverse of **P** can be written as

$$\mathbf{P}^{-1} = \begin{pmatrix} 1/x_1 & 0 & \ldots & 0 \\ -(x_2/x_1) & 1 & \ldots & 0 \\ \vdots & & & \\ -(x_n/x_1) & 0 & \ldots & 1 \end{pmatrix} \qquad (8.40)$$

The corresponding similarity transformation can be written for a 3 × 3 matrix as

$$\mathbf{P}^{-1}\mathbf{AP} = \begin{pmatrix} \lambda_1 & (a_{12}/x_1) & (a_{13}/x_1) \\ 0 & a_{22} - (x_2/x_1)a_{12} & a_{23} - (x_2/x_1)a_{13} \\ 0 & a_{32} - (x_3/x_1)a_{12} & a_{33} - (x_3/x_1)a_{13} \end{pmatrix} = \begin{pmatrix} \lambda_1 & b_{12} & b_{13} \\ 0 & & \\ 0 & \mathbf{B} & \end{pmatrix} \tag{8.41}$$

where variables b_{12}, b_{13}, and matrix $\mathbf{B}$ are defined by equating the two matrices. The characteristic determinant of the similarity transformation can be written as

$$|\mathbf{P}^{-1}\mathbf{AP} - \lambda\mathbf{I}| = (\lambda_1 - \lambda)|\mathbf{B} - \lambda I| \tag{8.42}$$

Similarly, the characteristic determinant of the right side of (8.41) can be written

$$\begin{vmatrix} \lambda_1 - \lambda & b_{12} & b_{13} \\ 0 & \mathbf{B} - \lambda\mathbf{I} & \\ 0 & & \end{vmatrix} = 0 \tag{8.43}$$

Expanding the determinant about the first column, we have

$$(\lambda_1 - \lambda)|\mathbf{B} - \lambda\mathbf{I}| = 0 \tag{8.44}$$

Equating (8.42) and (8.43), we get

$$|\mathbf{P}^{-1}\mathbf{AP} - \lambda\mathbf{I}| = (\lambda_1 - \lambda)|\mathbf{B} - \lambda\mathbf{I}| = 0 \tag{8.45}$$

Since the characteristic values of similar matrices are equal, the solution of (8.45) will also result in the characteristic value of matrix **A**. Thus, the second characteristic value of **A** can be obtained by solving for the characteristic value of matrix **B**.

Normalizing (8.41) so that $x_1 = 1$, we get

$$\mathbf{B} = \begin{pmatrix} a_{22} - x_2a_{12} & \ldots & a_{2n} - x_2a_{1n} \\ \ldots & & \\ a_{n2} - x_na_{12} & \ldots & a_{nn} - x_na_{1n} \end{pmatrix} \tag{8.46}$$

Calculate λ_2 by the iteration sequence $\mathbf{BY}_o, \ldots, \mathbf{B}^k\mathbf{Y}_o$, and so on. Having the value of λ_2, we may proceed to compute the second characteristic vector $\mathbf{X}_2$ of matrix **A**.

The similarity transformation matrix $\mathbf{P}^{-1}\mathbf{AP}$ will have characteristic vectors

$$\begin{pmatrix} y_1 \\ \mathbf{Y} \end{pmatrix}$$

We determine y_1 from the normalized ($x_1 = 1$) form of equation (8.41) as follows:

$$\begin{pmatrix} \lambda_1 & a_{12} & a_{13} \\ 0 & & \\ 0 & \mathbf{B} & \end{pmatrix} \begin{pmatrix} y_1 \\ \\ \mathbf{Y} \end{pmatrix} = \begin{pmatrix} \lambda_1 y_1 + a_{12}y_2 + a_{13}y_3 \\ \\ \mathbf{BY} \end{pmatrix} \tag{8.47}$$

$$= \lambda \begin{pmatrix} y_1 \\ \mathbf{Y} \end{pmatrix}$$

From the above relation

$$y_1 = \frac{a_{12}y_2 + a_{13}y_3}{\lambda - \lambda_1} \tag{8.48}$$

Consider the following relationships with **W** representing the characteristic vector of $\mathbf{P}^{-1}\mathbf{AP}$:

$$\mathbf{P}^{-1}\mathbf{AP}(\mathbf{W}) = \lambda(\mathbf{W}) \tag{8.49}$$

or

$$\mathbf{AP}(\mathbf{W}) = \lambda\mathbf{P}(\mathbf{W}) \tag{8.50}$$

Thus $\mathbf{P}(\mathbf{W})$ represents the characteristic vector of matrix **A**, by the definition of characteristic vectors. Finally, the second characteristic vector of **A**, $\mathbf{X}_2$, becomes

$$\mathbf{X}_2 = \mathbf{P}\begin{pmatrix} y_1 \\ \mathbf{Y} \end{pmatrix} \tag{8.51}$$

where **Y** is the characteristic vector of matrix **B**.

Example: As an example of the earlier procedure, consider the matrix

$$\mathbf{A} = \begin{pmatrix} -306 & -198 & 426 \\ 104 & 67 & -147 \\ -176 & -114 & 244 \end{pmatrix} \tag{8.52}$$

and assume we have calculated the first characteristic value and the corresponding vector,

$$\lambda_1 = 6\,; \quad \mathbf{X}_1 = \begin{pmatrix} 1 \\ -1/2 \\ 1/2 \end{pmatrix} \tag{8.53}$$

Note that the vector $\mathbf{X}_1$ is normalized with respect to the first element $x_1 = 1$. We then form matrix **B** of equation (8.46):

$$\mathbf{B} = \begin{pmatrix} -32 & 66 \\ -15 & 31 \end{pmatrix} \tag{8.54}$$

The characteristic values and the characteristic vectors of the **B** matrix are

$$\lambda_2 = -2 \quad \mathbf{Y} = \begin{pmatrix} 11 \\ 5 \end{pmatrix} \qquad \text{and} \quad \lambda_3 = 1 \quad \mathbf{Y} = \begin{pmatrix} 2 \\ 1 \end{pmatrix} \tag{8.55}$$

Consider the value of λ_2 and corresponding vector **Y**. Using equation (8.48), we get

$$y_1 = \frac{(-198)(11) + (426)(5)}{-2 - 6} = 6 \tag{8.56}$$

Using this and the elements of vector $\mathbf{X}_1$, we compute $\mathbf{X}_2$ from equation (8.51)

$$\mathbf{X}_2 = \begin{pmatrix} 6 \\ 8 \\ 8 \end{pmatrix} \text{ or } \begin{pmatrix} 3 \\ 4 \\ 4 \end{pmatrix} \tag{8.57}$$

Second Characteristic Value and Vector of Symmetric Matrices

Assume **X** and λ are the first calculated characteristic vector and value of matrix **A**. We want to find the second characteristic vector and value **Y** and μ. For symmetric matrices, the orthogonality condition

$$\mathbf{X} \cdot \mathbf{Y} = 0 \tag{8.58}$$

applies. Equation (8.58) states that the dot product of the characteristic vectors **X** and **Y** is zero.

We also have the defining equation of the second characteristic value and vector μ and **Y**,

$$\mathbf{AY} = \mu\mathbf{Y} \tag{8.59}$$

Denoting the components of **X** by $x_1, x_2, \ldots, x_n$, and the components of **Y** by $y_1, y_2, \ldots, y_n$, the orthogonality relation (8.58) can be written

$$x_1y_1 + x_2y_2 + \ldots + x_ny_n = 0$$

Solving for y_n, we get

$$y_n = -\frac{x_1}{x_n}y_1 - \frac{x_2}{x_n}y_2 - \ldots - \frac{x_{n-1}}{x_n}y_{n-1} \tag{8.61}$$

The relation in matrix form can be written

$$\begin{pmatrix} 1 & 0 & 0 & \ldots & 0 \\ 0 & 1 & 0 & \ldots & 0 \\ \ldots & & & & \\ -(x_1/x_n) & -(x_2/x_n) & -(x_3/x_n) & \ldots & 0 \end{pmatrix} \begin{pmatrix} y_1 \\ y_2 \\ \vdots \\ y_n \end{pmatrix} = \begin{pmatrix} y_1 \\ y_2 \\ \vdots \\ y_n \end{pmatrix} \tag{8.62}$$

Denoting (8.62) as

$$\mathbf{CY} = \mathbf{Y} \tag{8.63}$$

we get from

$$\mathbf{AY} = \mu\mathbf{Y} \tag{8.64}$$

the following

$$\mathbf{ACY} = \mu\mathbf{Y} \tag{8.65}$$

The iteration process can proceed to calculate μ and **Y**, using equation (8.65).

Having the values λ, **X**, and μ, **Y**, we can proceed to calculate the third characteristic value and vector ν and **Z** by using the orthogonality relations

$$\mathbf{X} \cdot \mathbf{Z} = 0$$

$$\mathbf{Y} \cdot \mathbf{Z} = 0 \tag{8.66}$$

Eliminating z_n and z_{n-1}, the components of vector **Z**, from the preceding equations, we get equations of the form

$$\begin{pmatrix} 1 & 0 & 0 & \ldots & 0 & 0 \\ 0 & 1 & 0 & \ldots & 0 & 0 \\ \ldots & & & & & \\ - & - & - & \ldots & 0 & 0 \\ - & - & - & \ldots & 0 & 0 \end{pmatrix} \begin{pmatrix} z_1 \\ z_2 \\ - \\ - \\ - \\ z_{n-1} \\ z_n \end{pmatrix} = \begin{pmatrix} z_1 \\ z_2 \\ - \\ - \\ - \\ z_{n-1} \\ z_n \end{pmatrix} \quad (8.67)$$

where dashes represent values. Denoting the large matrix **B**, we have

$$\mathbf{BZ} = \mathbf{Z} \quad (8.68)$$

Since **Z** and ν are the characteristic vector and value of matrix **A**, we also have $\mathbf{AZ} = \nu\mathbf{Z}$. Substituting for **Z** on the right of this equation from (8.68), we get

$$\mathbf{ABZ} = \nu\mathbf{Z} \quad (8.69)$$

We can follow the iteration process with (8.69) to obtain the characteristic value and vector ν and **Z** of matrix **A**.

Example: Consider the symmetrical matrix

$$\mathbf{A} = \begin{pmatrix} 2 & 1 \\ 1 & 3 \end{pmatrix} \quad (8.70)$$

It is desired to obtain the first and the second characteristic values and vectors of matrix **A**. For the initial iteration, assume

$$\mathbf{X} = \begin{pmatrix} 1 \\ 1 \end{pmatrix} \quad (8.71)$$

and proceed with iteration

$$\begin{pmatrix} 2 & 1 \\ 1 & 3 \end{pmatrix} \quad \begin{pmatrix} 1 \\ 1 \end{pmatrix} = \begin{pmatrix} 3 \\ 4 \end{pmatrix} = 4\begin{pmatrix} 0.75 \\ 1 \end{pmatrix}$$

$$\begin{pmatrix} 0.75 \\ 1 \end{pmatrix} = 3.75 \begin{pmatrix} 0.6666 \\ 1 \end{pmatrix}$$

$$\begin{pmatrix} 0.6666 \\ 1 \end{pmatrix} = 3.666 \begin{pmatrix} 0.6363 \\ 1 \end{pmatrix} \tag{8.72}$$

$$\begin{pmatrix} 0.6363 \\ 1 \end{pmatrix} = 3.6363 \begin{pmatrix} 0.6248 \\ 1 \end{pmatrix}$$

$$\begin{pmatrix} 2 & 1 \\ 1 & 3 \end{pmatrix} \begin{pmatrix} 0.6248 \\ 1 \end{pmatrix} = 3.6248 \begin{pmatrix} 0.6204 \\ 1 \end{pmatrix}$$

The last iteration results in

$$\lambda_1 = 3.6248\,; \quad \mathbf{X} = (0.6204,\ 1)' \tag{8.73}$$

The next characteristic value and vector are obtained by using equation (8.65) as follows:

$$\begin{pmatrix} 2 & 1 \\ 1 & 3 \end{pmatrix} \begin{pmatrix} 1 & 0 \\ -0.6204 & 0 \end{pmatrix} \begin{pmatrix} y_1 \\ y_2 \end{pmatrix} = \mu \begin{pmatrix} y_1 \\ y_2 \end{pmatrix}$$

$$\tag{8.74}$$

$$\begin{pmatrix} 1.379 & 0 \\ -0.8612 & 0 \end{pmatrix} \begin{pmatrix} y_1 \\ y_2 \end{pmatrix} = \mu \begin{pmatrix} y_1 \\ y_2 \end{pmatrix}$$

Start the iteration by $\mathbf{Y} = (1,1)'$ and calculate

$$\begin{pmatrix} 1.379 & 0 \\ -0.8612 & 0 \end{pmatrix} \begin{pmatrix} 1 \\ 1 \end{pmatrix} = \begin{pmatrix} 1.379 \\ -0.8612 \end{pmatrix} = 0.8612 \begin{pmatrix} 1.601 \\ -1 \end{pmatrix}$$

$$\tag{8.75}$$

$$\begin{pmatrix} 1.601 \\ -1 \end{pmatrix} = \begin{pmatrix} 2.207 \\ -1.378 \end{pmatrix} = 1.378 \begin{pmatrix} 1.601 \\ -1 \end{pmatrix}$$

We have, from the last values,

$$\mu = 1.378 \text{ and } \mathbf{Y} = (1.601, -1)' \tag{8.76}$$

Method of A.M. Danilevsky

The method of A.M. Danilevsky is developed for the calculation of the coefficients of the characteristic polynomial rather than characteristic values themselves. The characteristic values are obtained through the solution of the characteristic polynomial by using a root finding method (such as Lin-Bairstow's method[1]. Having obtained the characteristic values we can go back to Danilevsky's method to find the corresponding characteristic vectors.

Consider the matrix

$$\mathbf{A} = \begin{pmatrix} a_{11} & a_{12} & \dots & a_{1n} \\ a_{21} & a_{22} & \dots & a_{2n} \\ \dots & & & \\ a_{n1} & a_{n2} & \dots & a_{nn} \end{pmatrix} \tag{8.77}$$

We are required to find the characteristic equation of this matrix. The method of Danilevsky accomplishes this through a series of similarity transformations which leave the characteristic equation unaffected.

If we can reduce matrix **A** to the form

$$\mathbf{P} = \begin{pmatrix} P_1 & P_2 & \dots & P_{n-1} & P_n \\ 1 & 0 & \dots & 0 & 0 \\ 0 & 1 & \dots & 0 & 0 \\ \dots & & & & \\ 0 & 0 & \dots & 1 & 0 \end{pmatrix} \tag{8.78}$$

we can write the characteristic determinant and subsequently the characteristic equation as follows:

$$P(\lambda) = \begin{pmatrix} P_1 - \lambda & P_2 & \dots & P_{n-1} & P_n \\ 1 & -\lambda & \dots & 0 & 0 \\ 0 & 1 & \dots & - & 0 \\ 0 & 0 & \dots & 1 & -\lambda \end{pmatrix} \tag{8.79}$$

$$= (-1)^n(\lambda^n - P_1\lambda^{n-1} - \dots - P_n) = 0$$

Thus, we can obtain the characteristic equation by having the elements of matrix (8.78). We then solve for the root of this equation to obtain the characteristic values. Having obtained the characteristic values, we proceed to compute the corresponding vectors, as will be discussed.
Consider the transformation matrix

$$\mathbf{M}_{n-1} = \begin{pmatrix} 1 & 0 & \dots & 0 & 0 \\ 0 & 1 & \dots & 0 & 0 \\ \dots & & & & \\ -\dfrac{a_{n1}}{a_{n,n-1}} & -\dfrac{a_{n2}}{a_{n,n-1}} & \dots & \dfrac{1}{a_{n,n-1}} & -\dfrac{a_{nn}}{a_{n,n-1}} \\ 0 & 0 & & 0 & 1 \end{pmatrix} \tag{8.80}$$

with the inverse

$$\mathbf{M}_{n-1}^{-1} = \begin{pmatrix} 1 & 0 & \dots & 0 & 0 \\ 0 & 1 & \dots & 0 & 0 \\ \dots & & & & \\ a_{n1} & a_{n2} & \dots & a_{n,n-1} & a_{nn} \\ 0 & 0 & \dots & 0 & 1 \end{pmatrix} \tag{8.81}$$

Matrix **C** is written by using these matrices and the original matrix **A**

$$\mathbf{C} = \mathbf{M}_{n-1}^{-1}\mathbf{A}\mathbf{M}_{n-1}$$

$$\mathbf{C} = \begin{pmatrix} c_{11} & c_{12} & \dots & c_{1,n-1} & c_{1n} \\ c_{21} & c_{22} & \dots & c_{2,n-1} & c_{2n} \\ \dots & & & & \\ c_{n-1,1} & c_{n-1,2} & \dots & c_{n-1,n-1} & c_{n-1,n} \\ 0 & 0 & \dots & 1 & 0 \end{pmatrix} \tag{8.82}$$

Note that equation (8.82) is equivalent to performing a similarity transformation on matrix **A**. Matrix **B** is designated as follows:

$$\mathbf{B} = \mathbf{A}\mathbf{M}_{n-1} = \begin{pmatrix} b_{11} & \dots & b_{1,n-1} & b_{1n} \\ b_{21} & \dots & b_{2,n-1} & b_{2n} \\ \dots & & & \\ 0 & \dots & 1 & 0 \end{pmatrix} \tag{8.83}$$

where

$$b_{ik} = a_{ik} + a_{i,n-1}\, m_{n-1,k} \qquad \text{for all } i \leq n - 1;\ k \neq n-1$$

$$b_{i,n-1} = a_{i,n-1}\, m_{n-1,n-1} \tag{8.84}$$

where

$$m_{n-1,i} = -\frac{a_{ni}}{a_{n,n-1}} \qquad \text{for } i \neq n-1$$

$$m_{n-1,n-1} = \frac{1}{a_{n,n-1}} \tag{8.85}$$

or, finally,[b]

$$\mathbf{C} = \mathbf{M}_{n-1}^{-1}\mathbf{B} = \begin{bmatrix} b_{11} & b_{12} & \cdots & b_{1,\,n-1} & b_{1n} \\ b_{21} & b_{22} & \cdots & b_{2,\,n-1} & b_{2n} \\ \cdots & & & & \\ \sum_{k=1}^{n} a_{nk}b_{k1} & \sum_{k=1}^{n} a_{nk}b_{k2} & \cdots & \sum_{k=1}^{n} a_{nk}b_{k,n-1} & \sum_{k=1}^{n} a_{nk}b_{kn} \\ 0 & 0 & \cdots & 1 & 0 \end{bmatrix} \tag{8.86}$$

The process of similarity transformations of matrix **C** is continued until the form given by equation (8.78) is achieved. The number of multiplications and divisions by the method of Danilevsky is $(n - 1)(n^2 + n - 1) \simeq 0(n^3)$.

Characteristic vectors corresponding to each characteristic value, computed from (8.79), can be determined from similarity transformation matrices **M**.

Assume $\mathbf{X} = (x_1, x_2, x_3, \ldots, x_n)'$ represents the characteristic vector of **A** corresponding to λ, and **Y** represents the characteristic vector of **P** corresponding to λ. Then

$$\mathbf{X} = \mathbf{M}_{n-1}\mathbf{M}_{n-2} \ldots \mathbf{M}_2\mathbf{M}_1\mathbf{Y} \tag{8.87}$$

Thus, to compute **X** we need to compute **Y**, the characteristic vector of **P**. Denoting the components of **Y** by $y_1, y_2, \ldots, y_n$ and using (8.78), the following set of equations can be written:

---

[b] In case the element $c_{k,k-1}$ at the end of some iteration becomes zero, a nonzero element is chosen along the same row, $c_{ki}(i < k - 1)$ and the ith and $k - 1$ columns and then the ith and $k - 1$ rows are interchanged (see Hovanessian and Pipes, *Digital Computer Methods in Engineering*, p. 95).

$$
\begin{aligned}
(P_1 - \lambda)y_1 + P_2y_2 + \ldots + P_ny_n &= 0 \\
y_1 - \lambda y_2 \qquad\qquad &= 0 \\
\ldots \qquad\qquad & \\
y_{n-1} - \lambda y_n &= 0
\end{aligned}
\tag{8.88}
$$

Normalizing $y_n = 1$, we get

$$y_{n-1} = \lambda,\ y_{n-2} = \lambda^2,\ \ldots,\ y_1 = \lambda^{n-1} \tag{8.89}$$

Using these values in (8.87) will result in the characteristic vector of matrix **A**.

Example: As an example of the method of Danilevsky, consider the matrix

$$\mathbf{A} = \begin{bmatrix} 1 & 3 \\ 6 & 2 \end{bmatrix} \tag{8.90}$$

We calculate the matrices of similarity transformation as follows:

$$\mathbf{M} = \begin{bmatrix} 1/6 & -1/3 \\ 0 & 1 \end{bmatrix} \quad \mathbf{M}^{-1} = \begin{bmatrix} 6 & 2 \\ 0 & 1 \end{bmatrix} \tag{8.91}$$

Using these relations, we may compute the **C** matrix of equation (8.82) which, in this case, will be equal to the $P(\lambda)$ matrix of (8.79), resulting in

$$
\begin{aligned}
\mathbf{C} = \mathbf{M}^{-1}\mathbf{A}\mathbf{M} &= \begin{bmatrix} 6 & 2 \\ 0 & 1 \end{bmatrix} \begin{bmatrix} 1 & 3 \\ 6 & 2 \end{bmatrix} \begin{bmatrix} 1/6 & -1/3 \\ 0 & 1 \end{bmatrix} \\
\mathbf{P}(\lambda) &= \begin{bmatrix} 3 & 16 \\ 1 & 0 \end{bmatrix}
\end{aligned}
\tag{8.92}
$$

The characteristic equation then becomes,

$$
\begin{aligned}
P(\lambda) &= (-1)^n(\lambda^n - P_1\lambda^{n-1} - \ldots - P_n) = 0 \\
&= (-1)^2(\lambda^2 - 3\lambda - 16) = 0 \\
&\lambda^2 - 3\lambda - 16 = 0
\end{aligned}
\tag{8.93}
$$

The solution of equation (8.93) will result in characteristic values.

Jacobi's Method for Characteristic Values and Vectors

Jacobi's method is based upon a series of similarity transformations which use orthogonal matrices. This method, unlike the method of A.M. Danilevsky, applies to symmetrical matrices. The basis of Jacobi's method is covered in this theorem.

Theorem. For a real symmetric matrix **A**, all eigen values are real, and a real orthogonal matrix **O** exists so that $\mathbf{O}^{-1}\mathbf{A}\mathbf{O}$ is diagonal. Note that for an orthogonal matrix $\mathbf{O}^{-1} = \mathbf{O}'$, i.e., the transpose of the matrix is equal to its inverse.

Consider the orthogonal transformation matrix involving sines and cosines:

$$\mathbf{O} = \begin{pmatrix} \cos\phi & -\sin\phi \\ \sin\phi & \cos\phi \end{pmatrix} \tag{8.94}$$

with

$$\mathbf{O}' = \mathbf{O}^{-1} = \begin{pmatrix} \cos\phi & \sin\phi \\ -\sin\phi & \cos\phi \end{pmatrix} \tag{8.95}$$

The similarity transformation results in

$$\mathbf{D} = \mathbf{O}^{-1}\mathbf{A}\mathbf{O} = \begin{pmatrix} \cos\phi & \sin\phi \\ -\sin\phi & \cos\phi \end{pmatrix} \begin{pmatrix} a_{ii} & a_{ik} \\ a_{ik} & a_{kk} \end{pmatrix} \begin{pmatrix} \cos\phi & -\sin\phi \\ \sin\phi & \cos\phi \end{pmatrix} \tag{8.96}$$

where

$$\begin{aligned} |a_{ik}| &= \text{maximum for } i \neq k \text{ and } a_{ki} = a_{ik} \\ d_{ii} &= a_{ii}\cos^2\phi + 2a_{ik}\sin\phi\cos\phi + a_{kk}\sin^2\phi \\ d_{ik} = d_{ki} &= -(a_{ii} - a_{kk})\sin\phi\cos\phi + a_{ik}(\cos^2\phi - \sin^2\phi) \\ d_{kk} &= a_{ii}\sin^2\phi - 2a_{ik}\sin\phi\cos\phi + a_{kk}\cos^2\phi \end{aligned} \tag{8.97}$$

Choosing ϕ so that $d_{ik} = d_{ki} = 0 \therefore \tan 2\phi = [2a_{ik}/(a_{ii} - a_{kk})]$ and setting $R^2 = (a_{ii} - a_{kk})^2 + 4a_{ik}^2$, we get

$$\begin{cases} d_{ii} = \dfrac{1}{2}(a_{ii} + a_{kk} - R) \\ \\ d_{kk} = \dfrac{1}{2}(a_{ii} + a_{kk} - R) \\ \\ d_{ik} = d_{ki} = 0 \end{cases} \tag{8.98}$$

Also $d_{ii} + d_{kk} = a_{ii} + a_{kk}$ and $d_{ii}d_{kk} = a_{ii}a_{kk} - a_{ik}^2$

A series of two-dimensional orthogonal transformations having the form given in the elements (i,i), (i,k), (k,i) and (k,k) and identical with the unit matrix elsewhere will reduce matrix **A** to the diagonal form. The elements of this diagonal matrix will be equal to the characteristic values of **A**. The elements i and k are chosen so that $|a_{ik}|$ will be maximum with $i \neq k$. It can also be shown that the product of the orthogonal transformation $\mathbf{O}_1\mathbf{O}_2\mathbf{O}_3, \ldots, \mathbf{O}_r$ will result in the **O** matrix, the columns of which represent the characteristic vectors of **A**.

Example: Find the characteristic values of the symmetric matrix

$$\begin{pmatrix} 2 & -1 & 0 \\ -1 & 2 & -1 \\ 0 & -1 & 1 \end{pmatrix} \tag{8.99}$$

using Jacobi's method. The characteristic values of this matrix are 3.238, 1.555, and 0.198.[2]

Using Jacobi's transformation with $i = 2$ and $k = 3$, we get

$$a_{ii} = 2, \qquad a_{ki} = a_{ik} = -1; \qquad a_{kk} = 1 \tag{8.100}$$

and

$$\tan 2\phi = \frac{2a_{ik}}{(a_{ii} - a_{kk})} = \frac{2(-1)}{2 - 1} = -2 \tag{8.101}$$

resulting in

$$\phi = -31.7175 \text{ degrees}, \cos\phi = 0.8507 \text{ and } \sin\phi = -0.5257. \tag{8.102}$$

The transformation corresponding to equation

$$\begin{pmatrix} \cos\phi & \sin\phi \\ -\sin\phi & \cos\phi \end{pmatrix} (a_{ij}) \begin{pmatrix} \cos\phi & -\sin\phi \\ \sin\phi & \cos\phi \end{pmatrix} \tag{8.103}$$

becomes

$$\begin{pmatrix} 1 & 0 & 0 \\ 0 & 0.8507 & -0.5257 \\ 0 & 0.5257 & 0.8507 \end{pmatrix} \begin{pmatrix} 2 & -1 & 0 \\ -1 & 2 & -1 \\ 0 & -1 & 1 \end{pmatrix}$$

$$\begin{pmatrix} 1 & 0 & 0 \\ 0 & 0.8507 & 0.5257 \\ 0 & -0.5257 & 0.8507 \end{pmatrix} \tag{8.104}$$

The product results

$$\begin{pmatrix} 2 & -0.8507 & -0.5257 \\ -0.8507 & 2.6181 & 0 \\ -0.5257 & 0 & 0.3820 \end{pmatrix} \tag{8.105}$$

Using this matrix, let $i = 1$ and $k = 2$ with

$$a_{ii} = 2, \quad a_{ki} = a_{ik} = -0.8507, \quad a_{kk} = 2.6181 \tag{8.106}$$

and

$$\tan 2\phi = \frac{2(-0.8507)}{2 - 2.6181} \tag{8.107}$$

resulting in

$$\begin{aligned} \phi &= 35.0172 \text{ degrees} \\ \cos \phi &= 0.8190 \\ \sin \phi &= 0.5738 \end{aligned} \tag{8.108}$$

Using the above values, the transformation matrix becomes,

$$\begin{pmatrix} 0.8190 & 0.5738 & 0 \\ -0.5738 & 0.8190 & 0 \\ 0 & 0 & 1 \end{pmatrix} \begin{pmatrix} 2 & -0.8507 & -0.5257 \\ -0.8507 & 2.6181 & 0 \\ -0.5257 & 0 & 0.3820 \end{pmatrix}$$

$$\begin{pmatrix} 0.8190 & -0.5738 & 0 \\ 0.5738 & 0.8190 & 0 \\ 0 & 0 & 1 \end{pmatrix} \tag{8.109}$$

The product of the matrices becomes

$$\begin{pmatrix} 1.4039 & 0 & -0.4318 \\ 0 & 3.2142 & 0.3017 \\ -0.4305 & 0.3017 & 0.3820 \end{pmatrix} \tag{8.110}$$

With the (8.110) matrix, now use $i = 1$, $k = 3$ with

$$\begin{aligned} \phi &= -20.1005 \text{ degrees} \\ \sin\phi &= -0.3437 \\ \cos\phi &= 0.9391 \end{aligned} \tag{8.111}$$

Using the (8.111) values, the transformation matrix will be

$$\begin{pmatrix} 0.9391 & 0 & -0.3437 \\ 0 & 1 & 0 \\ 0.3437 & 0 & 0.9391 \end{pmatrix} \begin{pmatrix} 1.4039 & 0 & -0.4318 \\ 0 & 3.2142 & 0.3017 \\ -0.4318 & 0.3017 & -0.3820 \end{pmatrix}$$

$$\begin{pmatrix} 0.9391 & 0 & 0.3437 \\ 0 & 1 & 0 \\ 0.3437 & 0 & 0.9391 \end{pmatrix} \tag{8.112}$$

The product of these matrices will result in

$$\begin{pmatrix} 1.5620 & -0.1037 & 0 \\ -0.1037 & 3.2142 & 0.2833 \\ 0 & -0.2833 & 0.224 \end{pmatrix} \tag{8.113}$$

Now consider the rotation $i = 2$ and $k = 3$ with

$$\begin{aligned} \phi &= 5.3648 \text{ degrees} \\ \sin\phi &= 0.0935 \\ \cos\phi &= 0.9956 \end{aligned} \tag{8.114}$$

This transformation will result:

$$\begin{pmatrix} 1.562 & -0.1032 & 0.0097 \\ -0.1032 & 3.2408 & 0 \\ 0.0097 & 0 & 0.1956 \end{pmatrix} \tag{8.115}$$

Further rotations of the (8.115) matrix will reduce the values of off-diagonal elements, rendering more exact characteristic values. In many engineering calculations, the values along the diagonal may be considered sufficiently close to the exact values of 3.238, 1.555, and 0.198 as given previously.

Example: Consider the matrix

$$\mathbf{A} = \begin{pmatrix} 10 & 7 & 8 & 7 \\ 7 & 5 & 6 & 5 \\ 8 & 6 & 10 & 9 \\ 7 & 5 & 9 & 10 \end{pmatrix} \tag{8.116}$$

It is desired to obtain its characteristic values by Jacobi's method. Choosing $i = 3$, $k = 4$, we get $\tan 2\phi = 18/(10 - 10) = \infty$ and $\phi = 45$ degrees. After first rotation

$$\mathbf{A} = \begin{bmatrix} 10 & 7 & 15/\sqrt{2} & -1/\sqrt{2} \\ 7 & 5 & 11/\sqrt{2} & -1/\sqrt{2} \\ 15/\sqrt{2} & 11/\sqrt{2} & 19 & 0 \\ -1/\sqrt{2} & -1/\sqrt{2} & 0 & 1 \end{bmatrix} \tag{8.117}$$

Take $i = 1$ and $k = 3$ and $\tan 2\phi = (15/\sqrt{(2)})/(10 - 19)$; $\phi = 56.4949$ degrees, and so on. After 14 rotations, diagonal elements become

$$\lambda_1 = 0.010150, \quad \lambda_2 = 0.843110, \quad \lambda_3 = 3.85804$$

$$\lambda_4 = 30.288686 \tag{8.118}$$

Discussion of the Methods

The method most often used by engineers in the determination of characteristic values and vectors is the power method. The reason for this is twofold. First, in a great number of engineering problems only the highest (or the lowest) characteristic value and the corresponding vector are required and, second, the power method gives the characteristic value and the characteristic vector simultaneously. The iteration method, depending on the separation of the characteristic values, usually converges rapidly.

Although the method for the evaluation of the highest characteristic value was given in the text, similar reasoning can be used to compute the lowest characteristic value, as given in the following paragraphs. The characteristic equation of matrix **A** is given as

$$\mathbf{AS} = \lambda\mathbf{S} \tag{8.119}$$

where **S** is the characteristic vector and λ the corresponding value. Equation (8.119) can also be written

$$\frac{1}{\lambda}(\mathbf{S}) = \mathbf{A}^{-1}\mathbf{S} \tag{8.120}$$

Since λ represents the largest characteristic value, 1/λ will represent the smallest characteristic value. According to equation (8.120), the smallest characteristic value can be obtained by inverting the matrix **A** and applying the power method on this inverse matrix.

Other methods, such as the method of Danilevsky described in this chapter, give the characteristic equation whereby a root-finding method should be used to calculate the characteristic values.[2] These methods are usually applied where the characteristic values are complex numbers. Although a variation of the power method can also be applied in cases where complex roots arise, it is usually easier to find the complex roots from the characteristic equation. Fortunately, the matrices arising in most engineering applications are real and symmetric, giving rise to real characteristic roots. In cases where "damping force proportional to velocity" is involved in vibration problems, the mathematical formulations result in sets of equations with complex characteristic values. These equations can be solved by matrix iteration methods.[3]

The Jacobi method of orthogonal transformations to obtain the characteristic values and vectors of a matrix is very popular with almost all digital computer programmers. This method has the advantage of resulting in all (rather than only one) of the characteristic values at the end of the iteration and it converges reasonably fast. Its disadvantage lies in the fact that it applies only to symmetrical matrices.

Problems

1. Obtain the first four iterations of the characteristic values and vectors of the matrix

$$\mathbf{A} = \begin{bmatrix} 1 & -4 & 2 \\ -4 & 1 & -2 \\ 2 & -2 & -2 \end{bmatrix}$$

starting with initial vector (1, −1, 1)′.

2. Obtain the first four iterations of the characteristic values and vectors of the matrix

$$\mathbf{A} = \begin{bmatrix} 2 & 1 & 1 \\ 1 & 3 & 1 \\ 1 & 1 & 4 \end{bmatrix}$$

starting with the initial vector (1, 1, 1,)′.

3. Obtain the first three iterations of the characteristic values and vectors of the matrix

$$\mathbf{A} = \begin{bmatrix} 1 & 1 & 1 \\ 1 & 2 & 2 \\ 1 & 2 & 3 \end{bmatrix}$$

starting with the initial iteration (1, 1, 1)′.

4. After squaring the matrix of problem 1, find the first three iterations of the characteristic values and vectors. Start with an initial vector of (1, −1, 1)′. After three iterations, we will have $\mathbf{A}^8\mathbf{Y}_o$. Obtain $\mathbf{A}^9\mathbf{Y}_o$ and compute λ_1.
5. After squaring the matrix of problem 2, obtain the first three iterations of the characteristic values and vectors. Start with initial iteration of (1, 1, 1)′. After three iterations, we will have $\mathbf{A}^8\mathbf{Y}_o$. Obtain $\mathbf{A}^9\mathbf{Y}_o$ and compute λ_1.
6. Using the scalar product method, obtain the value of λ from the first three iterations of problem 1. Start with an initial vector of (1, −1, 1)′.
7. Using the scalar product method, obtain the value of λ from the first three iterations of problem 3. Start with an initial vector of (1, 1, 1)′.
8. By inverting the matrix of problem 1, calculate the lowest characteristic value and the corresponding vector. Obtain the first three iterations, assuming (1, 1, 1)′ for the initial vector.
9. By inverting the matrix of problem 3, calculate the lowest characteristic value and vector by iteration. Obtain the first three iterations, using (1, −1, 1)′ for the initial vector.
10. In problem 9, assume an initial vector of (1, 1, 1) and examine the result after the first eight iterations. Comparing the results of this problem with those of problem 9, we see the importance of the starting vector in reducing the number of iterations for a convergent solution.

11. Assuming the first characteristic value and vector of problem 1 to be

$$\lambda = 6.0$$

$$\mathbf{X} = \begin{pmatrix} 2 \\ -2 \\ 1 \end{pmatrix}$$

find the second and third characteristic values by the method of similarity transformation.

12. Consider problem 3 with the first characteristic value and vector given as

$$\lambda_1 = 5.04892; \quad \mathbf{X} = \begin{pmatrix} 0.445042 \\ 0.801938 \\ 1.0 \end{pmatrix}$$

Using the similarity transformation method, compute the second highest characteristic value and the corresponding vector.

13. Consider the matrix

$$\mathbf{A} = \begin{pmatrix} -32 & 66 \\ -15 & 31 \end{pmatrix}$$

with $\lambda_1 = -2$ and

$$\mathbf{X} = \begin{pmatrix} 11 \\ 5 \end{pmatrix}$$

Find the second characteristic value and the corresponding vector using the method of similarity transformation.

14. Consider the matrix of problem 2 and assume the first characteristic value and the corresponding vector have been found to be

$$\lambda_1 = 5.21433 \quad \text{and} \quad \mathbf{X} = \begin{pmatrix} 0.52543 \\ 0.688896 \\ 1.0 \end{pmatrix}$$

Using orthogonality relations, obtain the first three iterations of the second characteristic value and vector using $(1, 1, 1)'$ as the initial vector.

15. Consider the matrix of problem 3 and assume the first characteristic value and vector have been found to be

$$\lambda_1 = 5.04892; \quad \mathbf{X} = \begin{pmatrix} 0.445042 \\ 0.801938 \\ 1.0 \end{pmatrix}$$

Using orthogonality relations, obtain the first three iterations of the second characteristic value and vector using (1, 1, 1)′ as the initial vector.

16. Consider the matrix

$$\begin{bmatrix} 5 & -1 & -1 \\ 1 & 3 & 1 \\ -2 & 2 & 4 \end{bmatrix}$$

Compute the coefficients of the characteristic equation by the method of A.M. Danilevsky.

17. Consider the matrix

$$\begin{bmatrix} 78 & -60 & 15 \\ 150 & -117 & 30 \\ 200 & -160 & 43 \end{bmatrix}$$

Compute the characteristic equation by the method of A.M. Danilevsky.

18. Using the information given in the example problem of equation (8.99), compute the characteristic vectors of this matrix.

19. Diagonalize the matrix

$$\begin{bmatrix} 1 & 3 \\ 3 & 2 \end{bmatrix}$$

by Jacobi's method and obtain the characteristic vectors.

20. Find the characteristic values and vectors of the matrix

$$\begin{bmatrix} 3 & 8 \\ 8 & 12 \end{bmatrix}$$

using Jacobi's method.

21. Consider the symmetrical matrix

$$\begin{bmatrix} 2 & -1 & 0 \\ -1 & 2 & -1 \\ 0 & -1 & 1 \end{bmatrix}$$

and assume that the first and second characteristic values and vectors are computed as follows

$$\lambda = 3.247 \qquad \mathbf{X} = \begin{pmatrix} 1.802 \\ -2.247 \\ 1.0 \end{pmatrix}$$

$$\mu = 1.555 \qquad \mathbf{Y} = \begin{pmatrix} -1.247 \\ -0.555 \\ 1.0 \end{pmatrix}$$

Compute the third characteristic value ν and the corresponding vector **Z**, using the orthogonality relations.

Notes

1. S.A. Hovanessian and L.A. Pipes, *Digital Computer Methods in Engineering* (New York: McGraw-Hill, 1969), p. 134.
2. Ibid. p. 117.
3. L.A. Pipes and S.A. Hovenessian, *Matrix Computer Methods in Engineering* (New York: Wiley, 1969), chap. 8

Bibliography

Faddeeva, V.N. *Computational Methods of Linear Algebra*. New York: Dover, 1959.

Fröberg, C.E. *Numerical Analysis*. Reading, Mass.: Addison-Wesley, 1965.

Hovanessian, S.A., Pipes, L.A. *Digital Computer Methods in Engineering*. New York: McGraw-Hill, 1969

Pipes, L.A., Hovanessian, S.A. *Matrix Computer Methods in Engineering*. New York: Wiley, 1969

Todd, J. *Numerical Analysis*. New York: McGraw-Hill, 1962.

Answers to Problems

Chapter 1, pp. 27-32

1. $\mathbf{L} = \begin{pmatrix} 1 & 0 & 0 \\ 1.2 & 1 & 0 \\ 1.3 & 2.036 & 1 \end{pmatrix}$; $\mathbf{U} = \begin{pmatrix} 10 & 16 & -2 \\ 0 & -11.20 & 26.40 \\ 0 & 0 & -45.15 \end{pmatrix}$

2. $\mathbf{L} = \begin{pmatrix} 1 & 0 & 0 \\ 0.375 & 1 & 0 \\ 0.750 & -0.939 & 1 \end{pmatrix}$; $\mathbf{U} = \begin{pmatrix} 16 & 5 & -3 \\ 0 & 6.125 & 13.125 \\ 0 & 0 & 32.574 \end{pmatrix}$

3. $\mathbf{L} = \begin{pmatrix} 1 & 0 \\ 0.313 & 1 \end{pmatrix}$; $\mathbf{U} = \begin{pmatrix} 16 & 8 \\ 0 & -14.504 \end{pmatrix}$

   $\text{Inverse} = \begin{pmatrix} 0.052 & 0.035 \\ 0.022 & -0.069 \end{pmatrix}$

4. $\mathbf{L}^{-1} = \begin{pmatrix} 1 & 0 \\ -10.75 & 1 \end{pmatrix}$; $\mathbf{U}^{-1} = \begin{pmatrix} 0.125 & 0.0114 \\ 0 & -0.0076 \end{pmatrix}$

   $\text{Inverse} = \begin{pmatrix} 0.0025 & 0.0114 \\ 0.0817 & -0.0076 \end{pmatrix}$

5. $\begin{pmatrix} 0.0429 & 0.1143 \\ 0.1143 & -0.0286 \end{pmatrix}$

6. $\begin{pmatrix} -0.2015 & 0.0746 & 0.1194 \\ 0.0746 & -0.0647 & 0.0299 \\ 0.1194 & 0.0299 & -0.0523 \end{pmatrix}$

7. $\begin{pmatrix} 0.0027 & -0.1271 & 0.1465 \\ -0.1271 & -0.1547 & 0.2653 \\ 0.1465 & 0.2652 & -0.2403 \end{pmatrix}$

8. $\mathbf{X}^0 = \begin{pmatrix} 8 \\ 12.8333 \end{pmatrix}$; $\mathbf{X}^1 = \begin{pmatrix} 10.1393 \\ 12.1669 \end{pmatrix} \ldots \mathbf{X}^3 = \begin{pmatrix} 9.9985 \\ 11.9980 \end{pmatrix}$

9. $\mathbf{X}^0 = \begin{pmatrix} 6 \\ 3.1667 \end{pmatrix}$; $\mathbf{X}^1 = \begin{pmatrix} 7.0555 \\ 2.8331 \end{pmatrix} \ldots \mathbf{X}^3 = \begin{pmatrix} 6.9247 \\ 3.0124 \end{pmatrix}$

10. $\mathbf{X}^0 = \begin{pmatrix} 4.8333 \\ -5.3333 \\ -3.0 \end{pmatrix} \ldots \mathbf{X}^3 = \begin{pmatrix} 55.8961 \\ -51.4044 \\ -76.9442 \end{pmatrix}$

11. $\mathbf{X}^0 = \begin{pmatrix} 8 \\ 12.1667 \end{pmatrix} \ldots \mathbf{X}^3 = \begin{pmatrix} 10.0 \\ 12.0 \end{pmatrix}$

12. $\mathbf{X}^0 = \begin{pmatrix} 6 \\ 2.8333 \end{pmatrix} \ldots \mathbf{X}^3 = \begin{pmatrix} 6.9273 \\ 2.7818 \end{pmatrix}$

13. $\mathbf{X}^0 = \begin{pmatrix} 4.8333 \\ -3.7222 \\ -9.7037 \end{pmatrix} \ldots \mathbf{X}^3 = \begin{pmatrix} 275.1590 \\ -546.203 \\ -1276.1 \end{pmatrix}$

14. $\mathbf{X}^0 = \begin{pmatrix} 6 \\ -3.6154 \\ 2.4649 \end{pmatrix} \ldots \mathbf{X}^3 = \begin{pmatrix} 5.4052 \\ -1.8676 \\ 1.8021 \end{pmatrix}$

15. $x_1 = -1.0;\quad x_2 = 3.0;\quad x_3 = -2.0$
16. $x_1 = 6.0;\quad x_2 = 3.0;\quad x_3 = -1.0$
17. $x_1 = 1.0;\quad x_2 = -1.0$
18. $x_1 = 1.9998;\quad x_2 = 5.9998$
19. $x_1 = 0.9995;\quad x_2 = 2.0002;\quad x_3 = 3.0002$
20. $x_1 = 4.9194;\quad x_2 = -1.7322;\quad x_3 = 1.5977$
21. $x_1 = -1.0;\quad x_2 = 3.0;\quad x_3 = -2.0$

22. $\begin{pmatrix} 0.0483 & 0.1129 \\ 0.0325 & -0.2581 \end{pmatrix}$

23. $\begin{pmatrix} 2.6429 & -0.0713 & -0.7857 \\ 1.2857 & -0.1428 & -0.5715 \\ 0.9286 & -0.2142 & -0.3572 \end{pmatrix}$

24. $\begin{pmatrix} 0.3884 & -0.3719 & -0.4794 \\ -0.1405 & 0.1984 & 0.3223 \\ -0.1157 & 0.2810 & 0.2066 \end{pmatrix}$

25. $\mathbf{D}_2 = \begin{pmatrix} 2.0016 & -1.0011 & -0.0039 \\ -1.0007 & 2.0005 & -0.9983 \\ 0.0013 & -1.0009 & 0.9968 \end{pmatrix}$

26. $\mathbf{D}_2 = \begin{pmatrix} 0.64699 & -0.17647 & -0.11765 \\ -0.17647 & 0.41176 & -0.05882 \\ -0.11765 & -0.05882 & 0.29411 \end{pmatrix}$

27. $M = 4.5484;\quad N = 2.2235;\quad P = -3.6734$
28. $M = 2.3387;\quad N = 1.4884;\quad P = 1.7675$

Chapter 2,
pp. 51-55.

1. Given in the problem.
2. Given in the problem.
3.

i	x_i	y_i	∇y_i	$\nabla^2 y_i$	$\nabla^3 y_i$	$\nabla^4 y_i$	
1	0	0					$y_i = x_i^3$
2	1	1	1				
3	2	8	7	6			
4	3	27	19	12	6		
5	4	64	37	18	6	0	
6	5	125	61	24	6	0	
7	6	216	91	30	6	0	
8	7	343	127	36	6	0	
9	8	512	169	42	6	0	
10	9	729	217	48	6	0	

4.

i	x_i	y_i	Δy_i	$\Delta^2 y_i$	$\Delta^3 y_i$	$\Delta^4 y_i$	
1	0	3	2	12	12	0	$y_i = 2x_i^3 + 3$
2	1	5	14	24	12	0	
3	2	19	38	36	12	0	
4	3	57	74	48	12	0	
5	4	131	122	60	12	0	
6	5	253	182	72	12	0	
7	6	435	254	84	12		
8	7	689	338	96			
9	8	1027	434				
10	9	1461					

5.

i	x_i	y_i	$\mu\delta y_i$	$\delta^2 y_i$	$\mu\delta(\delta^2 y_i)$	
1	0	1				$y = 2x_i^2 + 1$
2	1	3	4	4		
3	2	9	8	4	0	
4	3	19	12	4	0	
5	4	33	16	4	0	
6	5	51	20	4	0	
7	6	73	24	4	0	
8	7	99	28	4	0	
9	8	129	32	4		
10	9	163				

6. $y_0 = 1; \quad y_1 = \frac{1}{2}[-1 + 12\Delta x(1 - \Delta x)]$

$$y_2 = \frac{1}{1 + \Delta x}[8\Delta x - y_1 - y_0(1 - \Delta x)]$$

$$y_{i+1} = \frac{1}{1 + \Delta x}[8(i\Delta x) - y_{i-1} - y_{i-2}(1 - \Delta x)] \quad i = 2, 3, 4, \ldots$$

7. $y_0 = 1; \quad y_1 = 1; \quad y_2 = y_1[2 - 3(\Delta x)^4] - y_0 + 6\Delta x^3;$

$$y_{i+1} = y_i[2 - 3i^2(\Delta x)^4] - y_{i-1} + 6i(\Delta x)^3 \quad i = 2, 3, 4, \ldots$$

8. $y_0\left(1 - \frac{\Delta x^2}{2}\right) - y_1 + y_2\left(1 + \frac{\Delta x^2}{2}\right) = (\Delta x)^3$

$$y_1\left(1 - \frac{2\Delta x^2}{2}\right) - y_2 + y_3\left(1 + \frac{2\Delta x^2}{2}\right) = (2\Delta x)^3$$

...

$$y_{i-1}\left(1 - \frac{i\Delta x^2}{2}\right) - y_i + y_{i+1}\left(1 + \frac{i\Delta x^2}{2}\right) = (i\Delta x)^3$$

...

$$y_{N-2}\left(1 - \frac{N-1}{2}\Delta x^2\right) - y_{N-1} + y_N\left(1 + \frac{N-1}{2}\Delta x^2\right)$$

$N - 1$ equations in $N - 1$ unknowns; $y_0 = 1, \ldots y_N = 10$ with

$$\Delta x = \frac{1}{N}$$

9. $-3y_0 + 4y_1 - y_2 = 12\Delta x$

...

$$y_{i-1}\left[1 - \frac{i^2\Delta x^3}{2}\right] - y_i + y_{i+1}\left[1 + \frac{i\Delta x^3}{2}\right] = 6\Delta x^2$$

...

$$y_{N-2} - 4y_{N-1} + 3y_N = 20\Delta x$$

10. Given in the problem.

11. $A = \frac{y_{i+2} - 2y_{i+1} + y_i}{2\Delta x^2}; \quad B = \frac{-y_{i+2} + 4y_{i+1} - 3y_i}{2\Delta x}$

$$C = y_i; \quad \text{Area} = \frac{\Delta x}{3}[y_{i+2} + 4y_{i+1} + y_i]$$

12. Area $= (4/3)\Delta x(2y_{i+2} - y_{i+1} + 2y_i)$
13. Given in the problem.
14. Given in the problem.
15. $\log_{10} 1.024 = 0.0103000$
16. Given in the problem.
17. $\log_{10} 1.024 = 0.0103000$
18. $\tan 0.55 = 0.61311$
19. From Table, $D^2 \tan x = 1.62292$
    From differentiation, $D^2 \tan x = 1.62963$
20. Area, trapezoidal rule $= 0.04799$
    Area, Simpson's 1/3 rule $= 0.04799$

Chapter 3,
pp. 83-87

1. $y(\text{at } x = 0.1) = 1.19933$; $y(\text{at } x = 0.2) = 1.39467$
2. $y(\text{at } x = 1.1) = 3.4170$; $y(\text{at } x = 1.2) = 3.5160$
3. $y'(\text{at } x = 1.1) = 2.4760$
4. $y(\text{at } x = 2.1) = 1.6072$; $y(\text{at } x = 2.3) = 2.8809$
5. $y(\text{at } x = 0.2) = 1.3970$; $y(\text{at } x = 0.3) = 1.5897$
6. $y(\text{at } x = 1.1) = 3.43$; $y(\text{at } x = 1.5) = 3.0639$
7. Problem 1

   $$z' = -7z - 2y + 16; \quad z = 2 \text{ at } x = 0$$

   $$y' = z; \quad y = 1 \text{ at } x = 0$$

   Problem 2

   $$y' = z; \quad z = 6 \text{ at } x = 1$$

   $$z' = -3z - 8y + 8x; \quad y = 3 \text{ at } x = 1$$

8. $x = 0.10, \quad y = 1.1372, \quad y' = 0.8584$
   $x = 0.20, \quad y = 1.1879, \quad y' = 0.2212$
   $x = 0.30, \quad y = 1.1909, \quad y' = -0.1231$
9. $x = 0.20, \quad y = 1.1824, \quad y' = 0.2511$
10. $$\begin{aligned} \dot{y} &= z \\ \dot{z} &= w \\ \dot{w} &= -3w - 6z + 7y + t^2 + 6 \end{aligned} \qquad \begin{cases} y = 1 \text{ at } t = 0 \\ z = 0 \text{ at } t = 0 \\ w = 6 \text{ at } t = 0 \end{cases}$$

11. Given in problem.

12. $y_4 = y_0 + \frac{4h}{3}(2y_1' - y_2' + 2y_3')$ Predictor

Compute

$y_4' = -3y_4 + 6x_4^2$

$y_4 = y_2 + \frac{h}{3}(y_2' + 4y_3' + y_4')$ Corrector

13. $y_4' = y_0' + \frac{4h}{3}(2y_1'' - y_2'' + 2y_3'')$ (A)

Compute

$y_4 = y_2 + \frac{h}{3}(y_2' + 4y_3' + y_4')$ (B)

Having this, recalculate y_4''

$y_4'' = -7y_4' - 2y_4 + 16$ (C)

and

$y_4' = y_2' + \frac{h}{3}(y_2'' + 4y_3'' + y_4'')$ Corrector (D)

Using this, recompute y_4 from (B). Proceed with y_5 . . . .

14. Proceed in order

$y_4' = y_0' + \frac{4h}{3}(2y_1'' - y_2'' + 2y_3'')$ Predictor

$y_4 = y_2 + \frac{h}{3}(y_2' + 4y_3' + y_4')$ (B)

$y_4'' = -3y_4' - 8y_4 + 3x_4$

$y_4' = y_2' + \frac{h}{3}(y_2'' + 4y_3'' + y_4'')$ Corrector

Using the above and (B), recompute y_4 and proceed to $y_5, y_6, \ldots$.

15. $x = 0.95522$ at $t = 0.3$
$x = 0.92091$ at $t = 0.4$
$x = 0.87741$ at $t = 0.5$

16. Using Taylor series, solve for y_1 and y_2 at $x_1 = 1 + \Delta x$ and $x_2 = 1 + 2\Delta x$ and obtain F_0, F_1, and F_2.

$$y_3 = 2y_2 + \frac{5\Delta x^2}{6} F_2 - y_1 + \frac{h^2}{12} F_1$$

$F_3 = x_3^2 + 3y_3$

Using this F_3, calculate y_3 from

$$y_3 - \frac{h^2}{12} F_3 = 2y_2 + \frac{5h^2}{6} F_2 - y_1 + \frac{h^2}{12} F_1$$

Calculate F_3 with this y_3 and continue to compute $y_4, y_5, \ldots$.

17. $y_{n+1}\left[1 + \frac{3\Delta x}{2}\right] + y_n[\Delta x^2 - 2] + y_{n-1}\left[1 - \frac{3\Delta x}{2}\right] = x_n^2 \Delta x^2$

Setting $n = 1, \ldots, (N - 1)$ will give us $N - 1$ equations in $N - 1$ unknowns

$y_1, y_2, \ldots, y_{N-1}$.

18. $y_1[-3.99] + y_2[2.05] = -1.83$

$y_1[1.95] + y_2[-3.99] + y_3[2.05] = 0.12$

...

$y_8[1.95] + y_9[-3.99] = -3.98$

19. $y_1[-1.94] + y_2[1.05] = -0.94999$

$y_1[0.95] + y_2[-1.94] + y_3[1.05] = 0.00008$

...

$y_8[0.95] + y_9[-1.94] + y_{10}[1.05] = 0.00512$

Boundary condition $y_8 - 4y_9 + 3y_{10} = 0.40$

10 equations in 10 unknowns $y_1 - y_{10}$

20. $-3y_0 + 4y_1 - y_2 = 2$

$y_0(0.25) + y_1(-1.75) + y_2(1.75) = 0.125$

$n = 2, 3, \ldots, 9$

$y_{n-1}(0.25) + y_n(-1.75) + y_{n+1}(1.75) = n(0.125)$

Last equation: $y_8 - 4y_9 + 3y_{10} = 6$

11 equations in 11 unknowns $y_0 - y_{10}$

21. $\dot{y} = z; \quad \dot{z} = x; \quad \dot{x} = -2t^2x - 2ty + 10$

$$\begin{pmatrix} y \\ z \\ x \end{pmatrix}_{n+1} = \begin{pmatrix} 1 & \Delta t & 0 \\ 0 & 1 & \Delta t \\ -2t_n\Delta t & 0 & -2t_n^2\,\Delta t + 1 \end{pmatrix}_n \begin{pmatrix} y \\ z \\ x \end{pmatrix}_n + \begin{pmatrix} 0 \\ 0 \\ 10\Delta t \end{pmatrix}$$

22. $x_1 = x; \quad x_2 = y; \quad x_3 = \dot{y}$

$$\begin{pmatrix} x_1 \\ x_2 \\ x_3 \end{pmatrix}_{\text{at } t=0.1} = \begin{pmatrix} 1 \\ 0.1 \\ 1.1 \end{pmatrix}; \quad \begin{pmatrix} x_1 \\ x_2 \\ x_3 \end{pmatrix}_{\text{at } t=0.3} = \begin{pmatrix} 1.2410 \\ 0.3300 \\ 1.3120 \end{pmatrix}$$

23. Assume $\delta_0 = \sigma_0$; $\delta_1 = \sigma_1$ and compute

$$\nu_1 = \sigma_1(K) + \sigma_0(K\,\Delta t - K)$$

$$\delta_2 = \sigma_2 - \nu_1$$

$$\nu_2 = \delta_2(K) + \delta_1(K\,\Delta t - K) + \nu_1(1 + \Delta t)$$

$$\delta_3 = \sigma_3 - \nu_2$$

$$\nu_3 = \delta_3(K) + \delta_2(K\,\Delta t - K) + \nu_2(1 + \Delta t)$$

and so on.

24. $\sigma_{n+1} = \sigma_n(1 - b\Delta t) + A\,\Delta t\epsilon_n$

$\beta_{n+1} = \beta_n(1 - a\,\Delta t) + K\,\Delta t\sigma_{n+1}$

and also $\epsilon_1 = \rho_1 - \beta_1$

25. $x(nT) = C_1q^n + C_2q_2^n$ (nonoscillatory)

$x(nT) = q_1^n(C_1 + C_2n)$ (nonoscillatory)

$x(nT) = R^n(C_1' \cos n\theta + C_2' \sin n\theta)$ (oscillatory)

$\omega T < 2$ to get oscillatory solution.

Chapter 4,
pp. 115-119

1. 65.7 percent
2. 1.56 percent
3. 0.64 percent
4. 10.29 percent

5. $\mu = 95.5$; $\sigma = 21.16$
6. $\mu = 7.0185$; $\sigma = 4.0665$
7. $\mu = 6$; $\sigma = 5.84259$
8. Form a table of the following kind:

i	A_i	f_{o_i}	$m = \frac{\bar{A}_i - 95.5}{\sigma}$	$\psi(m)$	$f_{e_i} = \frac{(120)(\Delta A)\psi(m)}{\sigma}$	χ_i^2
1	45 to 55	3	2.15	0.03955	2.243	0.255
2	55 to 65	5	1.68	0.09728	5.517	0.048
...						
10	135 to 145	5	2.10	0.044	2.490	2.530
Samples: 120						Σ4.44

$\mu = 95.5$; $\sigma = 21.16$; $\bar{A}_i = 50, 60, \ldots$.
The probability of distribution being Gaussian is about 90 percent.

9. Form a table of the following kind:

i	A_i	f_{o_i}	$e^{-\bar{A}_i/\mu}$	ΔA_i	$\frac{162}{\mu} e^{-\bar{A}_i/\mu}$	$\Delta A_i \chi^2$
0	0 to 2.5	0	0.836	2.5	48.2	48.2
1	2.5 to 3.5	20	0.652	1	15	1.67
2	3.5 to 4.5	23	0.565	1	13	7.69
...						
12	23.5 to 46.5	1	0.0068	23	3.68	1.95
Samples: 162						Σ122.84

$\bar{A}_i = 1.25, 3, 4, 5, \ldots, 15, 20, 35$

$\mu = 7.02$

The probability of data being from an exponential distribution is almost *zero*.

10. Probability of Gaussian fit > 90 percent.
11. Probability of Gaussian fit > 90 percent.
12. Probability of exponential distribution < 10 percent.
13. Greater than 98 percent.
14. $56.71 \leq x_{M+1} \leq 134.28$

    $66.23 \leq x_{M+1} \leq 124.76$

15. $83 \leq \mu \leq 108$

    $91 \leq \mu \leq 100$

16. Given in the problem.

17. $59.96 \leq \mu \leq 142$

    $65.21 \leq \mu \leq 127.4$

18. $932.6 \leq \mu \leq 1961.6$

    $1100 \leq \mu \leq 1607$

19. $\chi^2(90,200) = 174.84$

20. $\chi^2(70,145) = 135.63$

    $\chi^2(30,145) = 153.65$

Chapter 5,
pp. 147-152

1. $A = 4.681$; $B = -2.480$
2. $A = 0.2839$; $B = 1.2301$; $C = 2.2467$
3. $A = -0.0357$; $B = 2.8570$; $C = -1.9071$

4. $n = 1 \begin{cases} \hat{x}_1 = 10 \\ \dot{x}_1 = 0 \\ x_{p_2} = 10 \end{cases}$ $n = 2 \begin{cases} \hat{x}_2 = 10.15 \\ \dot{x}_2 = 0.0125 \\ x_{p_3} = 10.20 \end{cases}$

   $\ldots$ $n = 5 \begin{cases} \hat{x}_5 = 10.8928 \\ x_5 = 0.0477 \\ x_{p_5} = 11.0834 \end{cases}$ $\ldots$

   $n = 9 \begin{cases} x_9 = 12.2282 \\ \dot{x}_9 = 0.0736 \\ x_{p_{10}} = 12.5228 \end{cases}$

5. $n = 1 \begin{cases} \hat{x}_1 = 10 \\ \dot{x}_1 = 0 \\ \ddot{x}_1 = 0 \\ x_{p_2} = 10 \end{cases}$ $n = 2 \begin{cases} 10.15 \\ 0.0031 \\ 0.0031 \\ 10.1875 \end{cases}$

   $\ldots$ $n = 9 \begin{cases} 13.1796 \\ 0.2353 \\ -0.0035 \\ 14.0929 \end{cases}$

6. Given in the problem.
7. Given in the problem.

8. Given in the problem.
9. Given in the problem.
10. Given in the problem.
11. Given in the problem.

12. $\hat{\mathbf{X}}_1 = \begin{pmatrix} 39925 \\ -150 \\ 0 \end{pmatrix}; \quad \mathbf{P}_1 = \begin{pmatrix} 30100 & 5100 & 50 \\ 5100 & 2600 & 50 \\ 50 & 50 & 50 \end{pmatrix}$

13. $\hat{\mathbf{X}}_2 = \begin{pmatrix} 37963.9 \\ -190.934 \\ -0.05706 \end{pmatrix}; \quad \mathbf{K}_2 = \begin{pmatrix} 0.57632 \\ 0.07418 \\ 0.000713 \end{pmatrix}$

$$\mathbf{P}_2 = \begin{pmatrix} 38416.7 & 7903.1 & 221.184 \\ 7903.14 & 2614.27 & 146.291 \\ 221.184 & 146.291 & 74.9643 \end{pmatrix}$$

14. $\hat{\mathbf{X}}_1 = \begin{pmatrix} 10 \\ 0 \\ 0 \end{pmatrix}; \quad \mathbf{K}_1 = \begin{pmatrix} 0.5 \\ 0 \\ 0 \end{pmatrix};$

$$\mathbf{P}_1 = \begin{pmatrix} 0.031664 & 0.000432 & 0.000008 \\ 0.000432 & 0.000216 & 0.000004 \\ 0.000008 & 0.000004 & 0.000002 \end{pmatrix}$$

$$\hat{\mathbf{X}}_2 = \begin{pmatrix} 10.2334 \\ 3.23733\ \mathrm{E}-3 \\ 5.58160\ \mathrm{E}-5 \end{pmatrix}; \quad \mathbf{K}_2 = \begin{pmatrix} 0.466845 \\ 6.47466\ \mathrm{E}-3 \\ 1.11632\ \mathrm{E}-4 \end{pmatrix}$$

$$\mathbf{P}_2 = \begin{pmatrix} 3.34692\ \mathrm{E}-2 & 1.27138\ \mathrm{E}-3 & 3.62652\ \mathrm{E}-5 \\ 1.27138\ \mathrm{E}-3 & 3.76996\ \mathrm{E}-4 & 1.19482\ \mathrm{E}-5 \\ 3.62652\ \mathrm{E}-5 & 1.19482\ \mathrm{E}-5 & 2.99911\ \mathrm{E}-6 \end{pmatrix}$$

15. $\hat{\mathbf{X}}_1 = \begin{pmatrix} 48500 & 0 & 0 \\ -100 & 0 & 0 \\ 10 & 0 & 0 \end{pmatrix}$

$$\mathbf{P}_1 = \begin{pmatrix} 254883 & 43448.3 & 4000 \\ 43448.3 & 8444.83 & 800 \\ 4000 & 800 & 105 \end{pmatrix}$$

$$\hat{\mathbf{X}}_2 = \begin{pmatrix} 46115.9 & 0 & 0 \\ -186.042 & 0 & 0 \\ 1.37439 & 0 & 0 \end{pmatrix}$$

$$\mathbf{P}_2 = \begin{pmatrix} 162468 & 20660.2 & 1514.83 \\ 20660.2 & 3403.21 & 271.755 \\ 1514.83 & 271.755 & 49.5748 \end{pmatrix}$$

Chapter 6,
pp. 179-180

1. $f(t) = 3.2423 \sin \omega t - 0.3602 \sin 3\omega t + 0.1297 \sin 5\omega t \ldots$
2. $f(t) = 3.28419 \sin \omega_1 t - 0.4050 \sin \omega_2 t + 0.1808 \sin \omega_5 t \ldots$
3. $W^0 = 1; \quad W^1 = 0.707 - 0.707j; \quad W^2 = -j; \quad \ldots;$

   $W^7 = 0.707 + 0.707j; \quad W^8 = 1$

4.

Column → 0 1 2 3 4 5 6 7 0

Row

$$
\begin{array}{c}0\\1\\2\\3\\4\\5\\6\\7\end{array}
\begin{bmatrix} G_0 \\ G_1 \\ G_2 \\ G_3 \\ G_4 \\ G_5 \\ G_6 \\ G_7 \end{bmatrix}
=
\begin{bmatrix}
W^{0(0,0)} & W^{0(0,1)} & W^{0(0,2)} & W^{0(0,3)} & W^{0(0,4)} & W^{0(0,5)} & W^{0(0,6)} & W^{0(0,7)} \\
W^{0(1,0)} & W^{1(1,1)} & W^{2(1,2)} & W^{3(1,3)} & W^{4(1,4)} & W^{5(1,5)} & W^{6(1,6)} & W^{7(1,7)} \\
W^{0(2,0)} & W^{2(2,1)} & W^{4(2,2)} & W^{2(2,3)} & W^{6(2,4)} & W^{2(2,5)} & W^{4(2,6)} & W^{6(2,7)} \\
W^{0(3,0)} & W^{3(3,1)} & W^{6(3,2)} & W^{1(3,3)} & W^{4(3,4)} & W^{7(3,5)} & W^{2(3,6)} & W^{5(3,7)} \\
W^{0(4,0)} & W^{4(4,1)} & W^{0(4,2)} & W^{4(4,3)} & W^{0(4,4)} & W^{4(4,5)} & W^{0(4,6)} & W^{4(4,7)} \\
W^{0(5,0)} & W^{5(5,1)} & W^{2(5,2)} & W^{7(5,3)} & W^{4(5,4)} & W^{1(5,5)} & W^{6(5,6)} & W^{3(5,7)} \\
W^{0(6,0)} & W^{6(6,1)} & W^{4(6,2)} & W^{2(6,3)} & W^{0(6,4)} & W^{6(6,5)} & W^{4(6,6)} & W^{2(6,7)} \\
W^{0(7,0)} & W^{7(7,1)} & W^{6(7,2)} & W^{5(7,3)} & W^{4(7,4)} & W^{3(7,5)} & W^{2(7,6)} & W^{4(7,7)}
\end{bmatrix}
\begin{bmatrix} F_0 \\ F_1 \\ F_2 \\ F_3 \\ F_4 \\ F_5 \\ F_6 \\ F_7 \end{bmatrix}
$$

Column →0 1 2 7

$$G_0 = F_0 + W^0 F_1 + W^0 F_2 + \ldots + W^0 F_7$$

$$G_1 = F_0 + W^1 F_1 + W^2 F_2 + \ldots + W^7 F_7$$

$$G_2 = F_0 + W^2 F_1 + W^4 F_2 + \ldots + W^6 F_7$$

$$G_3 = F_0 + W^3 F_1 + W^6 F_2 + \ldots + W^5 F_7$$

$$G_4 = F_0 + W^4 F_1 + W^0 F_2 + \ldots + W^4 F_7$$

$$G_5 = F_0 + W^5 F_1 + W^2 F_2 + \ldots + W^3 F_7$$

$$G_6 = F_0 + W^6 F_1 + W^4 F_2 + \ldots + W^2 F_7$$

$$G_7 = F_0 + W^7 F_1 + W^6 F_2 + \ldots + W^4 F_7$$

5. Given in the problem.
6. 0, 8, 4, 12, 2, 10, . . ., 3, 11, 7, 15
7. 12
8. 24
9. Solid from $k = 7$, $l = 1$; dashed from $k = 3$, $l = 1$
10. Row 18, column 0; row 2, column 0
11. $G_{n_1(2)+n_0(1)} = F_0 + W\exp[n_1(2) + n_0(1)]F_1 + W\exp[n_0(2)]F_2$

$$+ W\exp[n_0(2) + n_1(2) + n_0(1)]F_3$$

12. $g(0.1) = 0.02$; $g(0.2) = 0.066$; . . . $g(0.50) = 0.416$

Chapter 7,
pp. 199-201

1. For minimum $2\lambda > 0$; for maximum $2\lambda < 0$
2. Minimum $\lambda > -1/3$; maximum, $\lambda < -1/3$
3. $8x + 3y + 6\lambda_1 x + 3\lambda_2 = 0$

   $3x + 14y + \lambda_1 + 2\lambda_2 y = 0$

   $3x^2 + y = 0$

   $3x + y^2 = 0$
4. $x_1 = 6$; $x_2 = 4$; $F = 22$
5. $x_1 = 4.235$; $x_2 = 4.117$
6. $x_1 = 3$; $x_2 = 0$; $x_3 = 1$; $x_4 = 3$
7. $x_1 = 3$; $x_2 = 0$
8.

x_1	x_2	λ_1	λ_2	λ_3	λ_4	RHS
2	2	−2	1	−1		0
2		−1	1		−1	0
−2	−1					−3
1	1					16
		3	−16			$F + L$

9.

x_1	x_2	λ_1	λ_2	λ_3	λ_4	RHS
2	1	3	−1	−1		0
1	6	1	−1		−1	0
3	1					16
−1	−1					−4
		−16	4			$F + L$

10. $$\begin{aligned} \lambda_1 + 2\lambda_2 + 4\lambda_3 + 6\lambda_4 + 9\lambda_5 + 10\lambda_6 &= x \\ \lambda_1 + 4\lambda_2 + 6\lambda_3 + 7\lambda_4 + 6\lambda_5 + 5\lambda_6 &= y \\ \lambda_1 + \lambda_2 + \lambda_3 + \lambda_4 + \lambda_5 + \lambda_6 &= 1 \end{aligned}$$

11. $A = 1$; $B = 1$; $C = -1$

12. $$\begin{aligned} \lambda_1 + 2\lambda_2 + \lambda_3 &= x \\ \lambda_1 + 2\lambda_2 + 2\lambda_3 &= y \\ \lambda_1 + 3\lambda_2 + 2\lambda_3 &= z \\ \lambda_1 + \lambda_2 + \lambda_3 &= 1 \end{aligned}$$

13. $$\begin{aligned} \lambda_1 + 2\lambda_2 + \lambda_3 &= x \\ \lambda_1 + 2\lambda_2 + 2\lambda_3 &= y \\ \lambda_1 + 3\lambda_2 + 2\lambda_3 &= z \\ \lambda_1 + \lambda_2 + \lambda_3 &= 1 \end{aligned}$$

14. Given in the problem.

Chapter 8,
pp. 226-230

1. $7 \begin{pmatrix} 1.89286 \\ -1.89286 \\ 1.0 \end{pmatrix}$

2. $5.26705 \begin{pmatrix} 0.533981 \\ 0.704423 \\ 1.0 \end{pmatrix}$

3. $5.06452 \begin{pmatrix} 0.44586 \\ 0.802548 \\ 1.0 \end{pmatrix}$

4. $\mathbf{A}^9\mathbf{Y}_0 = 5.8888\begin{pmatrix} 2.01415 \\ -2.01415 \\ 1.0 \end{pmatrix}$

5. $\mathbf{A}^9\mathbf{Y}_0 = 5.21955\begin{pmatrix} 0.52623 \\ 0.69054 \\ 1.0 \end{pmatrix}$

6. $\lambda = \left(\dfrac{Y_3 \cdot Y_3}{Y_2 \cdot Y_3}\right)(4.5) = 6.02256$

7. $\lambda = \left(\dfrac{Y_3 \cdot Y_3}{Y_2 \cdot Y_3}\right)(5.06452) = 5.04890$

8. $-0.318182\begin{pmatrix} 0.857143 \\ 1.42857 \\ 1.0 \end{pmatrix}; \quad \lambda = -(1/0.318182) = -3.14285$

9. $3.16667\begin{pmatrix} 1.73684 \\ -2.21053 \\ 1.0 \end{pmatrix}; \quad \lambda = (1/3.16667) = 0.31579$

10. Iteration 7 $\quad 3.34848\begin{pmatrix} 1.88688 \\ -2.26923 \\ 1.0 \end{pmatrix}$

    Iteration 9 $\quad 3.29412\begin{pmatrix} 1.84203 \\ -2.29412 \\ 1.0 \end{pmatrix}$

    $\lambda = \dfrac{1}{3.29412} = 0.30356$

11. $\lambda_2 = -3; \quad \lambda_3 = -3$
12. $\lambda_2 = 0.643105; \quad$ vector $= (-1.246947, -0.554979, 1.0)'$
13. $\lambda_2 = 1; \quad$ vector $= (2,1)'$
14. After three iterations $\quad 2.42729\begin{pmatrix} -0.446184 \\ -1.11128 \\ 1.0 \end{pmatrix}$

15. After three iterations $\quad 0.630463\begin{pmatrix} -1.1908 \\ -0.586132 \\ 1.0 \end{pmatrix}$

16. $(4 - \lambda)(\lambda^2 - 8\lambda + 12) = 0$

17. $(3 - \lambda)(\lambda^2 - \lambda - 6) = 0$

18. $\begin{pmatrix} 0.7690 & -0.5738 & 0.2815 \\ 0.2766 & 0.69672 & 0.6614 \\ -0.5773 & -0.4305 & 0.6952 \end{pmatrix}$

19. $\Lambda = \begin{pmatrix} -1.54138 & 0 \\ 0 & 4.5414 \end{pmatrix}; \quad \begin{pmatrix} 0.76302 & 0.64637 \\ -0.64637 & 0.76302 \end{pmatrix}$

20. $\Lambda = \begin{pmatrix} 1.6786 & 0 \\ 0 & 16.677 \end{pmatrix}; \quad \begin{pmatrix} 0.8632 & 0.5048 \\ -0.5048 & 0.8632 \end{pmatrix}$

21. $\lambda_3 = 0.198; \quad \begin{pmatrix} 0.445 \\ 0.802 \\ 1.0 \end{pmatrix}$

Index

Index

About the Author

S.A. Hovanessian, a senior scientist at Hughes Aircraft Company, is a specialist in mathematical modeling of engineering problems for digital computer solution, numerical analysis, optimization methods, and cybernetics. Since obtaining the Ph.D. in applied mathematics and engineering from UCLA in 1958, Dr. Hovanessian has worked on a wide variety of problems, including oil-field recovery methods, economic studies involving mathematical programming, industrial process control, radar detecting and tracking systems, missile flight trajectory simulations, and cost-effectiveness and reliability studies. In addition to the present text, he is the author of three other books—*Matrix Computer Methods*, (with L.A. Pipes, Wiley, 1969), *Digital Computer Methods* (with L.A. Pipes, McGraw-Hill, 1969), and *Radar Detection and Tracking Systems* (Artech, 1973)—and of numerous technical articles. These publications are primarily the result of his teaching activities at UCLA, USC, and other universities. Dr. Hovanessian is a member of IEEE, ASME, Tau Beta Pi, and Sigma Xi. He holds professional registration in both electrical and mechanical engineering in the state of California.